STARGAZING FOR ALL

A guide to astronomy suitable for
both beginners and those more experienced

KAREN HEDGES

First edition printed and published in the United Kingdom 2025.

A CIP catalogue record of this book is available from the British Library.

ISBN (Paperback): 978-1-7391236-1-1
Imprint: Karen Hedges
Typesetting: Matthew J Bird

For further information about this book, please contact the author:
www.karenhedges.co.uk

For Royston

STARGAZING FOR ALL

A guide to astronomy suitable for
both beginners and those more experienced

CONTENTS

PART 1

GETTING STARTED

INTRODUCTION

My aim with this book is to guide you in enjoying the stars and the Moon in a straightforward manner that gets you stargazing in an instant, as well as introducing other topics of particular interest. Most of the objects I describe in this book can be viewed with the naked eye or binoculars. When they are viewed through a telescope the image will be different. There are different types of telescopes and the key point to bear in mind is that with most of them the image is upside-down. There are also plenty of apps and maps to help you find out what you are looking at and how to find objects, while the purpose of this book is to suggest and explain some main or easier-to-find objects of interest.

Light pollution is a major factor to contend with when viewing stars which is why I focus on the brightest and most readily identifiable stars. When using a smartphone app, do make sure you have the brightness set to its lowest setting and to use a red torch wherever possible. A head torch is a great idea as that leaves your hands free to hold a phone or map.

Have fun and wishing you *clear skies*!

Karen
April 2025

TOP TEN TIPS FOR OBSERVING

1.	Wrap up warmly. It gets **very cold** when standing still and especially during the better viewing month of February.	
2.	*Orion!*	Learn some constellations, just one or two at a time to make it a manageable and enjoyable experience.
3.	Check the seeing conditions: minimal atmospheric disturbance as well as clear skies.	
4.		Use a red light. Red light does not affect your night vision, i.e. your eyes adapt to the dark more readily as well as being able to see the stars more clearly.

5.	Use a notebook and pencil (for recording what you see and for sketches). Make a note of the date, time, and weather conditions.
6.	Use a **planisphere**. It will also tell you where and when the planets will be visible. Available to buy from many retailers and observatories.
7.	Learn to use naked eyes and averted vision. Averted vision means looking slightly to the side of the object while concentrating on the object. Requires practise! *The blinking nebula, NGC6826, is an object that most dramatically demonstrates averted vision. Stare directly at this blue-green planetary nebula and you see only the dim central star. Look slightly to the side and the faint nebula around the star appears suddenly. When you switch from straight on to averted vision, the nebula appears to blink on and off.*
8.	Use binoculars, for quick and easy viewing. If possible, rest them on a window ledge or use a tripod to have a steady view.
9.	Use a telescope, for longer term and more detailed observing. There is a lot of choice with a wide price range available.
10.	*Have fun! This is most important, and, on those occasions when finding an object or trying to see something more clearly is proving a challenge, be patient and your patience will be richly rewarded.*

THE ZODIAC

Do you glance at the '*Stars*' in the press? Is good fortune coming your way? The zodiac is the zone along the ecliptic, the Sun's apparent path across the sky throughout the year, which encompasses the constellations lying in that region. Familiar as the twelve-star signs in newspapers and magazines, there are actually *thirteen* constellations because the Sun also passes through part of Ophiuchus, the serpent holder.

Although Ophiuchus is also a constellation along the ecliptic, it is not a *sign* of the zodiac. The 12 signs of the zodiac represent 30 equal degree divisions of the sky, whereas the actual 13 constellations of the zodiac are of various sizes.

The official constellation borders were agreed upon by the International Astronomical Union (IAU) in the early 20th century. Originally, when the zodiacal star signs were so identified, it is thought that perhaps the fainter stars of Ophiuchus were not worthy of recognition. The Sun stays in each zodiacal constellation for around a month.

Ophiuchus is not without note though. Within that constellation, the bright white star Rasalhague is worth spotting. It is one of the early stars to appear and has a companion. Rasalhague, just north of the main star Antares in Scorpius, rotates so fast it almost tears itself to pieces. At 48 light years away, it is also one of the nearest stars to our Sun.

There is an old rhyme to help remember the constellations: *The ram, the bull, the heavenly twins, next to the crab the lion shines, the virgin and the scales, the scorpion, archer and he-goat, the man that pours the water out, and the fish with glittering scales.*

There are many stories associated with the stars and learning the ones associated with the zodiac helps explain the order of them, with 'zodiac' meaning 'circle of small animals'.

EARLY STARS

A small selection of easier to spot astronomical objects to whet the appetite...

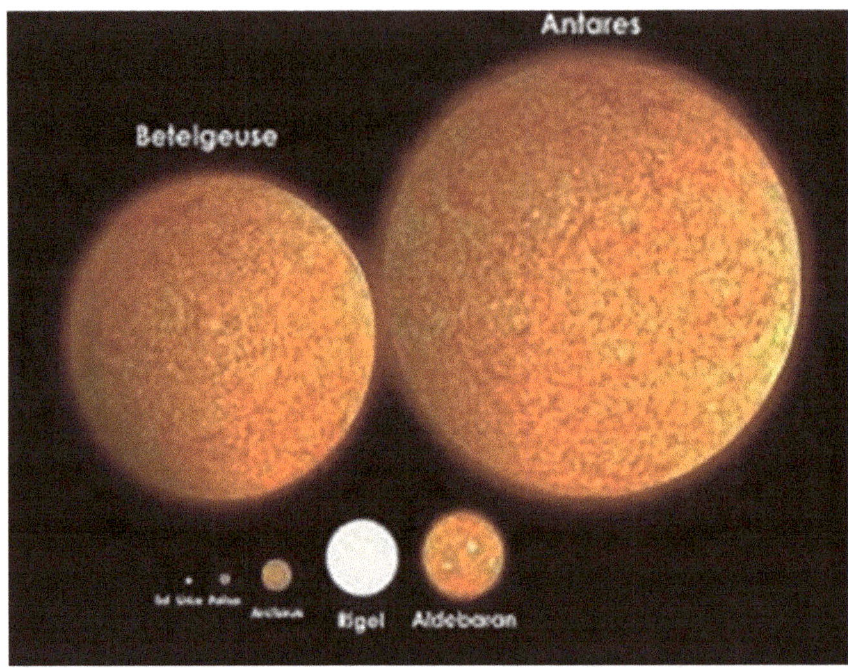

The Sun is the tiniest of the stars in this diagram.

The Sun

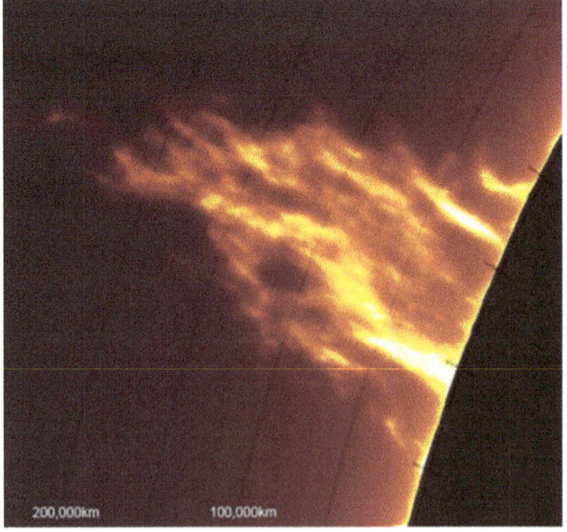

DO NOT UNDER ANY CIRCUMSTANCES LOOK AT THE SUN WITH THE NAKED EYE, OR THROUGH BINOCULARS, OR A TELESCOPE, WITHOUT SPECIAL EQUIPMENT.

An often-over-looked star, the Sun is an interesting object to visit an observatory for. With the aid of special solar filters, it is possible to view solar flares and sunspots (See Sun section page 140). I mention the Sun here because there are some activities you may safely enjoy. A sunrise or sunset are excellent starting points for observing its position and time.

Sunspots are simply cooler regions on the surface of the Sun and can be viewed safely when directing the sunlight onto a piece of paper through a piece of cardboard. The average sunspot is about the size of Earth.

Through a special filter solar flares can be seen, which are caused by the Sun sending out magnetic material (coronal mass ejection) reaching Earth, sometimes causing mayhem particularly in northern USA and Canada.

Aurorae (the Northern and Southern Lights)

Aurorae are the result of charged particles leaving the Sun and being borne towards the Earth on the solar wind. These are usually deflected by the Earth's magnetic field but, because this is weaker at the poles, the charged particles interact with the Earth's magnetic field giving rise to the splendid sight of aurorae. The diagram below shows the particles interacting with the magnetic field behind the Earth before being deflected to the poles where they appear as aurorae.

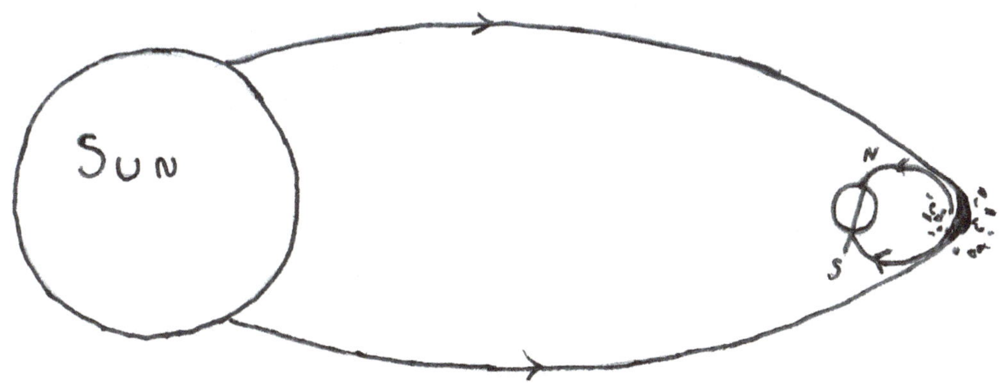

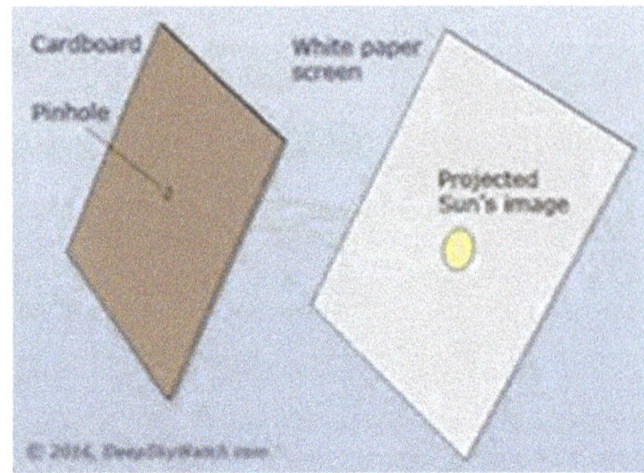

Venus

Although not a star, this planet is referred to as the Evening or Morning Star because of its very bright appearance. It is safe to look at this one. Again, a filter is needed when using equipment because the atmosphere is so thick it reflects sunlight making it difficult to see the planet itself. Sometimes, a crescent Venus can be seen, for the same reason we see phases of the Moon, i.e. due to the position of the planet in relation to the Sun. A pretty sight to look out for is the Moon and Venus together.

Polaris

The pole star is found overhead in the constellation of Ursa Minor. It is an important navigational aid as it appears stationary while the entire northern sky moves around it. Located nearly at the north celestial pole, it is the point around which the entire northern sky turns. Movement of the stars will take it closer to true pole position in 2100. It is actually a triple star system with two yellow-white companion stars.

Sirius

This star is clearly visible, appearing relatively low towards the southern horizon. Sirius comes from the Greek for 'scorching' or 'sparkling' due to its extra-twinkly nature. This white star is hot with a surface temperature of 9,940°K. It is known as the Dog Star because it is the brightest star in the constellation Canis Major (the greater dog), which is one of the two dogs belonging to Orion. The other dog is nearby, Canis Minor with Procyon the lead star. At a distance of only 8.7 light years away (or 80 million million kilometres or 50 million million miles), Sirius is one of the closest stars to the Sun. It is the brightest star in the sky with a companion star, Sirius B, which orbits it every 50 years. This companion star is bright in its own right but somewhat overshadowed by its neighbour.

The Plough

This well-known asterism is a subset of Ursa Major and known as the Plough, Panhandle or Big Dipper. It is clearly visible all year round. The two stars at the handle end act as pointers to the Pole Star.

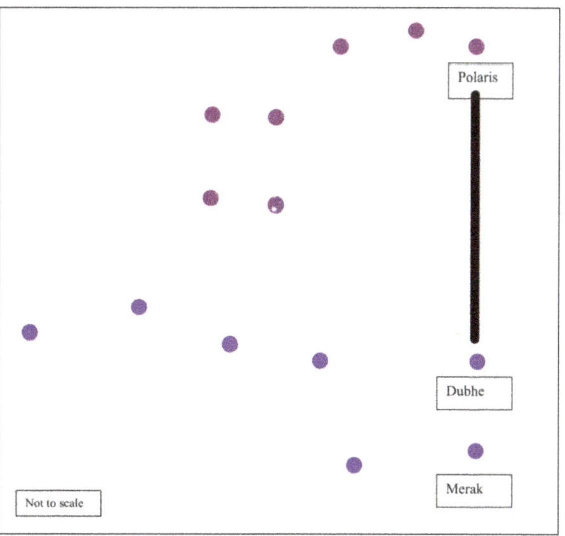

Arcturus

Using the handle of the Plough, one can arc down to Arcturus which is an orange giant, 37 light years away. Arcturus is in the constellation Boötes.

Spica

And from Arcturus one then takes a spike down to Spica… which is in the constellation of Virgo, harbinger of spring. Spica is a binary system, one of which is around 11 times the diameter of the Sun, and the other just 6 times.

Summer Triangle

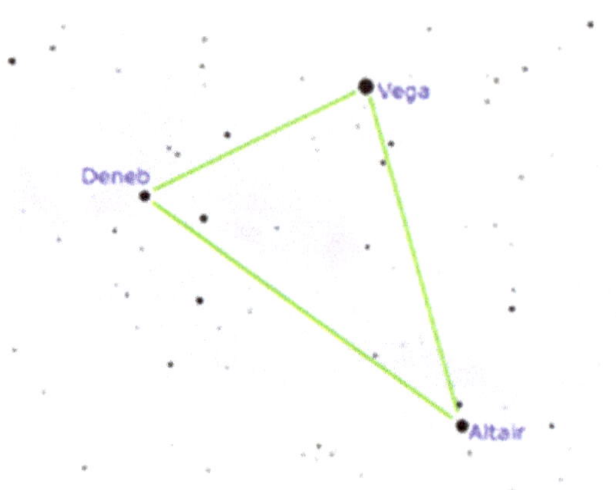

This is an **asterism**, or pattern of stars, made up of **Vega, Deneb** and **Altair** each leading the way to a different constellation, full details of which are featured in the section about constellations.

Winter Hexagon

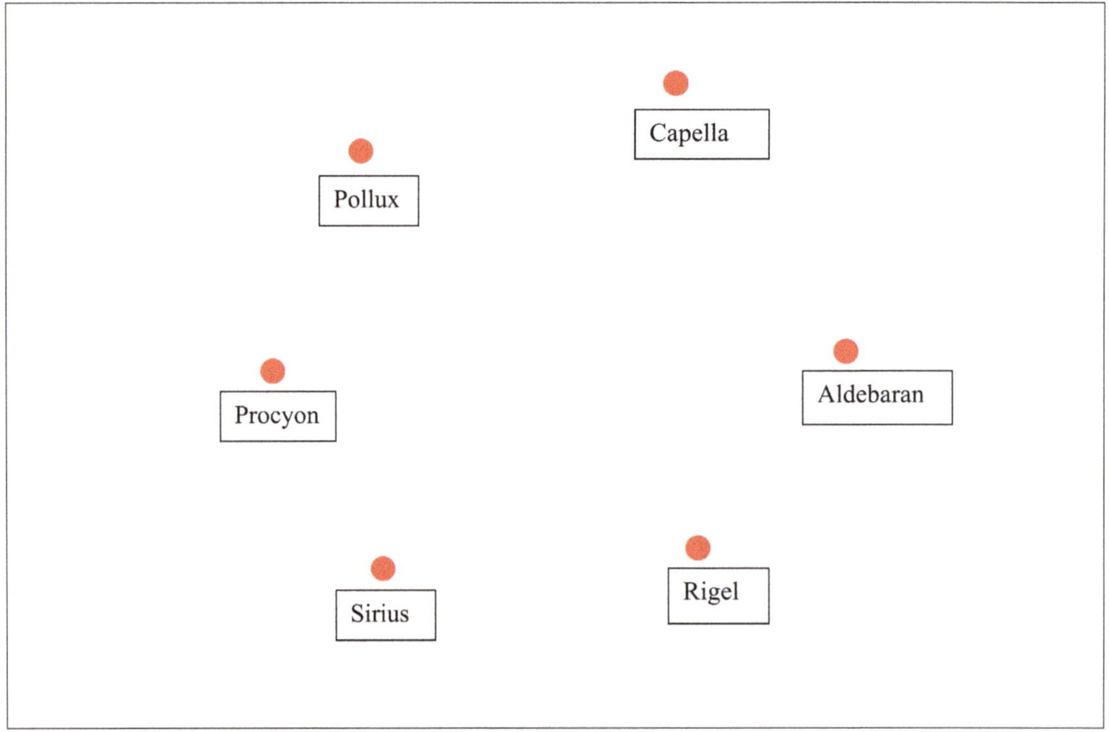

I really don't like doing this... but, again, there is a full section about this feature and its associated constellations later in the book. For now, the main thing is to try to spot the stars that make up the 'Hexagon': Sirius, Procyon, Pollux, Capella, Aldebaran, and Rigel.

Albireo in Cygnus

A great favourite. It can be seen as two stars, one blue, one white, through decent binoculars. At the head of the Swan, a great cross in the sky, it is relatively easy to find and enjoy.

Aldebaran

Although known as the red eye of the Bull, it is not really in Taurus as it is in front of the Hyades cluster which are 153 light years away while Aldebaran is 'only' 65 light years away; halfway between us and the Hyades. If you locate the Pleiades, then locating this red star will be straightforward as it is to the left of them. You will also be able to locate the Hyades through finding Aldebaran as they form the background to this star.

Alpha Vulpeculae

Near the head of Cygnus, 297 light years away, this is a pretty double star visible through binoculars. It is an orange giant with a companion 200 light years away. The faint constellation of Vulpecula within which it sits, is interesting because it contains the first pulsar (*pulsating radio star*) to be seen.

Altair

Within the constellation of Aquila, this white star is 17 light years away, making it one of the closest to us.

Antares

Although low on the horizon in the northern hemisphere, it is worth trying to locate this star as it is one of the massive ones in the universe and is the main star of the constellation Scorpius.

Barnard's Star

A faint star six light years away, which holds the honour of being the star which moves at the fastest rate. It moves across one degree of sky per every 350 years. It moves so quickly that in a human lifetime it can change position by half the diameter of the Full Moon across our sky in relation to the distant stars. Lying within the constellation of Ophiuchus, this star was discovered by US astronomer E. E. Barnard.

Betelgeuse

A red Supergiant, 1000 times larger than the Sun and 59,000 times as bright as the Sun: if it was where our Sun is, it would extend out as far as Jupiter! It is 640 light years away and is the brightest star within the constellation of Orion.

Capella

The brightest star (or two… it is a binary) in the constellation, Auriga, Capella is overhead in the winter whereas Vega is overhead in the summer months. Both stars in its binary system are around 3 times the mass of the Sun. These two stars are a brilliant illustration of how the Earth tilts to and away from the Sun over time. In summer, we are tilting towards Vega while in winter, we tilt towards Capella. One of the stars in Auriga is shared by Taurus, so two constellations for the price of one star can be located through Capella.

Cassiopeia and Andromeda Galaxy

A big 'W' to find in the sky. This one acts as a location finder for the mighty Andromeda Galaxy, which is visible to the naked eye as a fuzzy blob. The galaxy is below the second 'V' of the 'W'. If you are able to locate the fainter strands of Andromeda, the galaxy is towards the middle and to the left of the left-hand line of stars.

Deneb

A White supergiant, it is the most distant bright star at 3,000 light years away within the constellation of Cygnus (part of the Summer Triangle).

Fomalhaut

The star Fomalhaut within the constellation of the Southern Fish, is sometimes called the Autumn Star by people in the Northern Hemisphere, while it's a spring star south of the equator. It's famous in astronomical science as the first star with a visible exoplanet. It appears in a part of the sky that's largely empty of bright stars. For this reason, Fomalhaut is often called the Lonely One or Solitary One. It's an easy star to spot and one you'll want to see. It is a white ordinary star, sometimes appearing red due to atmospheric disturbance due to its position low on the horizon in the northern hemisphere.

Garnet Star

Or Mu Cephei, to give it is formal name, in the constellation of Cepheus, is 1,650 times diameter of Sun and 100,000 times as bright and 2,800 light years away. It is a variable red supergiant carbon star. The name was given by William Herschel, the astronomer who discovered Uranus, due to its fiery hue. This is quite an easy star to locate. I managed to find this red star using a little star book and no fancy apps on an evening when others were struggling even with fancy apps on their phones. This star is a Standard Candle, i.e. a star whose known brightness is used to determine the brightness of other stars.

Great Square of Pegasus

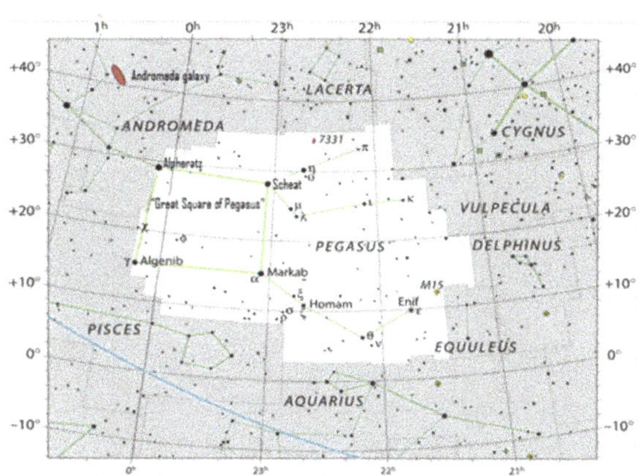

This is an asterism or pattern of stars forming the body of the horse in the constellation of Pegasus (the winged horse in Greek mythology), which is located below Cassiopeia, and to the left of Cygnus. The square comprises four stars: Alpheratz, a blue-white star at the head of the Andromeda constellation, with the other three within Pegasus: Algenib which is 9 times the mass of the Sun, Scheat, an old red giant and Markab, another old giant star. These four stars frame a fairly empty square of sky. However, it is a feature I have struggled with at times because I could not work out which square of stars was this one…

Pleiades – a group of 7 relatively easy stars to spot…

Through binoculars, you will be rewarded by the sight of a glorious array of many stars. The Pleiades are also known as the Seven Sisters (although there are a quite a few more than 7 though – around 500!) as you can generally see 7 stars with the naked eye. This star cluster is one of the brightest in the sky, about 444 light years from Earth. From side to side the group spans 13 light years, or about halfway from Earth to the bright star of Vega. Like a school of fish, the stars move together as a gravitationally bound swarm through space. An example of an open cluster of stars, the group continues to make new stars, thus replacing those that move away from the cluster over a period of time, i.e. hundreds of millions of years.

Procyon

The bright star in Canis Minor, its name means 'before the dog' as it appears before Sirius. It is among the brightest stars in the sky and comprises two stars in a binary system. Part of the reason for its brightness is its closeness to Earth at just 12 light years away.

Ras Algethi (Alpha Herculis)

(with Rasalhague nearby to the east)

Red supergiant around 400 times the diameter of the Sun, with an emerald-green secondary star. It is about 360 light years away and apart from its companion (another double) by 500 AU. Meaning 'kneeler's head' Ras Algethi is found at the head of the constellation of Hercules.

Rasalhague

Spotting this star leads you to the constellation of Ophiuchus which is noteworthy for being omitted from the signs of the Zodiac even though it is on the path of the ecliptic with the other 'signs'. Ophiuchus represents a serpent, and the main star means 'head of the serpent' in Arabic. The star rotates so fast it almost tears itself to pieces. It is one of the nearest stars to the Sun at just 48 light years away.

Regulus

The main star in Leo, Regulus is a blue-white star, part of a multiple star system consisting of at least four stars. This bright star system is 79 light years from

Earth. Two of the stars are in a binary system taking millions of years to orbit each other. One of the stars rotates every 15 hours.

Rigel

This blue-white supergiant is a double star 777 light years away, and 51,000 times as bright as the Sun. It is the second brightest star in Orion. The temperature of this hot blue star is 12,300°C.

T Lyrae

This is found in the southwest corner of Lyra (to the right of Vega). It is 2,064 light years away and is visible with 10x50 binoculars. Although this star appears somewhat faint because of this distance, it appears very luminous as there are not many other stars around it. With a magnitude of +7 it is a very bright object in any event.

Vega

This star is the brightest of the three stars which make up the Summer Triangle and is almost directly overhead during summer. The brightness is due in part to its proximity to Earth and the fact that it is 52 times as bright as the Sun. It is one of the nearest stars to us at a mere 25 light years away. Like our Sun, it is halfway through its lifespan. At one time it was the pole star and will be again due to the movement of the stars (precession). The star is about 450 million years old. This brilliant white star is the fifth brightest in the entire sky.

The brightness or magnitude of stars is measured according to how bright they appear in the sky from our perspective on Earth. Magnitude measures the difference in brightness between stars with Vega being 0 as it is upon this star the scale is based; first magnitude being the brightest as seen from Earth down to sixth apparent magnitude.

Circumpolar constellations

These are the constellations that surround the Pole Star (Polaris) and are therefore visible at all times of the year. These constellations are **Auriga, Camelopardalis, Cassiopeia, Cepheus, Draco, Lynx, Perseus, Ursa Major,** and **Ursa Minor**. Ursa Major, containing the Plough, is a well-known constellation. I was delighted to catch a glimpse of the fainter Camelopardalis one dark summer evening.

Plaskett's Star

Included here more for fun than as an actual viewing target… although you might catch a glimpse of this distant star, 6600 light years away, in the constellation of Monoceros. At 100 times the mass of the Sun it is one of the most massive stars in the universe. This star illustrates the scale of the universe as it appears just as a pinprick of light due to the distance from Earth. It is sobering, and interesting, to contemplate how big it would be the nearer one got to it.

The above selection of stars will give you an indication of the variety of stars in the universe as well as providing food for thought in terms of time and distance and the immense scale of the universe with its stars ranging from a star just slightly larger than the planet Jupiter, to those like Plaskett's Star.

THE MOON

This image was taken by an amateur, Royston Williams, using a DSLR and just zooming in and then in again.

Introduction to the Earth's Moon

In a lot of books, the Sun is shown as shining brightly by day and the Moon by night... but actually the Moon can be seen during the daytime as well. This makes it a good object for starting to learn about astronomy or practice observation techniques as you don't need to wait for dark evenings for general observation of its movement, phases, and location. Full Moon, first and last quarter phases are best for catching the Moon in the daytime (you can check in the press and online for dates and times of these). If the Moon is near the Sun, do not be tempted to use equipment in case you inadvertently look at the Sun and become blinded.

A daytime moon (Royston Williams)

The name, Moon, derives from an Old English word for month, 'moonth', which was used to refer to the Moon. Its regular cycle has long been part of human knowledge. The Earth's moon is referred to as *The* Moon because it was the only moon known about until 1610, when Galileo discovered some moons orbiting Jupiter. The Moon then gave its name to objects orbiting other planets. Interestingly, the word 'calendar' comes from the Latin for 'onset of the month'.

The current thinking about how the Moon was formed is that there was a collision between a large ball of rock (about the size of Mars, so smaller than Earth) and the Earth itself. A large chunk of the Earth's crust was knocked off and the loose bits of debris came together over time to form the Moon (most planets and moons were formed in this way in the early days of the Universe).

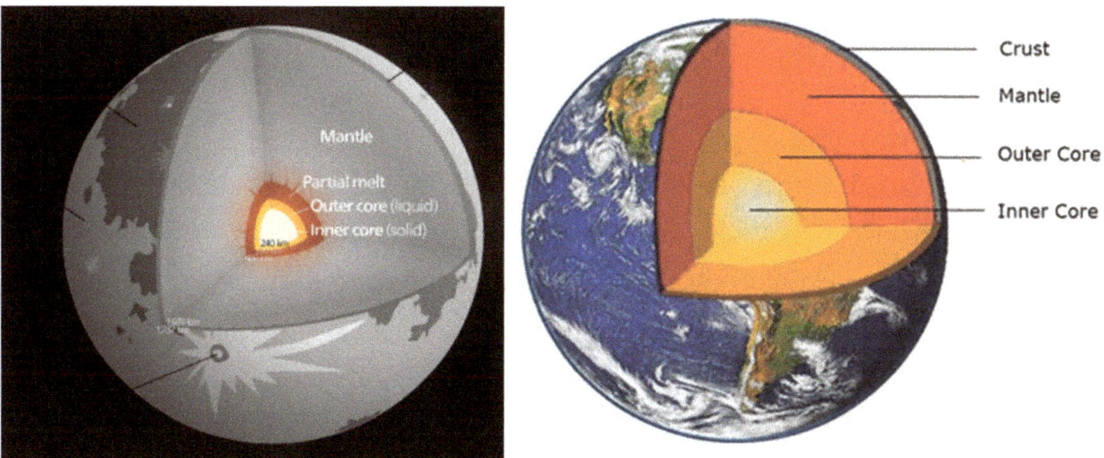

Let's look at Earth's composition first, the crust of the Earth makes up about 1% of its total volume i.e. 8 km/5 miles thick, while the core takes up around 50% of the volume of the Earth. By contrast, the Moon has a smaller core of just 20% of its volume with a 1300 km/806 miles thick outer mantle with a thin layer of regolith on top. This 9m/29-feet layer of regolith is made up of a mixture of powdery dust and bits of rock.

Neither the Earth nor the Moon are completely round but slightly squashed (*oblate spheres)* due to gravitational pull. The friction slows the rotation of the Earth down by 2.3 milliseconds every hundred years. This also has the effect of the Moon moving away from the Earth at just over 3cm a year.

The Earth's moon is the largest moon in proportion to its neighbouring planet in the solar system. The Moon has a wealth of features to enjoy, but simply observing its whereabouts in the sky can provide endless hours of enjoyment. If you get into the habit of taking a notebook and pencil with you, you can make quick sketches of its location relative to the Earth and Sun as well as making sketches of any features you see. Sketching is a useful tool to aiding closer observation as it focuses the mind on the detail rather than just enjoying the view. (*It is ok to just enjoy the view*!)

Here are some ideas to get you started using just your eyes:

Full Moon

Make a note of the time, date and where it is in relation to where you have observed it from. This will come in handy when you begin to understand its cycles and to compare your notes at the same time each month/year/event. This applies of course to *all* observations, not just the Full Moon. The Full Moon is relatively easy to spot which makes this point in its cycle a good starting point for logging its position.

Phases of the Moon

Phases are a result of the relative positions of the Sun, Earth and Moon and describe how much of the Moon's disc is illuminated from our perspective on Earth. The Moon, like Earth, is a sphere, and it is always half-illuminated by the Sun. From Earth we see more or less of only one side or face of the Moon as it travels around the Earth due its rotation and orbit... more of which later, and it is parts of this half of the Moon which we see in light or shadow according to the position of the Moon and Earth in relation to the Sun.

New Moon: The Moon is between Earth and the Sun, and the side of the Moon facing toward us receives no direct sunlight; it is lit only by dim sunlight reflected from Earth.

Just after New Moon: At New Moon, the Moon is not visible but soon a very thin crescent can be seen. You could try to spot the very first eyelash-thin edge as this happens. It is stunning. This is a beautiful sight in a clear sky and exciting to glimpse.

Waxing crescent: As the Moon moves around Earth, the side we can see gradually becomes more illuminated (waxing) by direct sunlight.

First quarter: The Moon is 90 degrees away from the Sun in the sky; it has travelled about a quarter of the way around the Earth since New Moon. Half of the face towards us illuminated from our viewpoint. The corresponding quarter of the lit half is on the far side of the moon which we cannot see.

Waxing gibbous: The area of illumination continues to increase or wax (gibbous means bulging). More than half of the Moon's face that is toward us is now lit by the Sun.

Full Moon: The Moon is at the exact opposite point in the sky from the Sun and the side facing Earth is fully illuminated. When there is a lunar eclipse, the Earth blocks the sunlight reaching whole or part of the Moon.

Waning gibbous: More than half of the Moon's face is illuminated, but the amount is decreasing (waning).

Last quarter: The Moon has moved another quarter of the way around Earth, to the third quarter position. The corresponding quarter of the lit half is on the far side of the moon which we cannot see, as in first quarter.

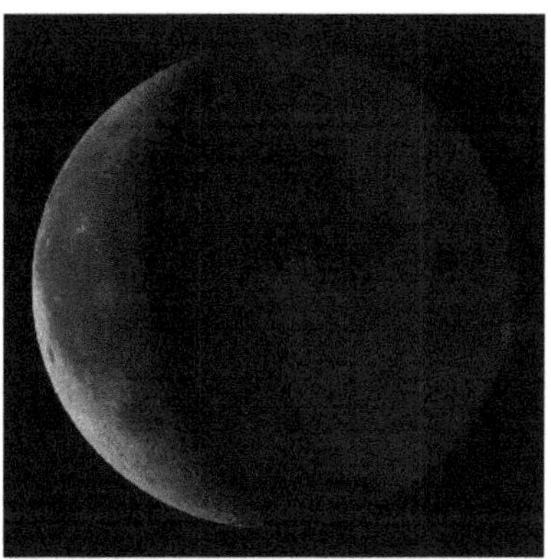

Waning crescent: Less than half of the Moon's face is illuminated, and the amount is decreasing. And then it starts all over again!

Earthshine: When the Moon is about 4 days into its cycle (4 days old), it stays long enough above the horizon to be seen against a fairly dark sky. This is often referred to as 'the old moon in the new moon's arms'. This effect is caused by light reflected back from the Earth and onto the Moon, called Earthshine. Sometimes you can glimpse the Moon faintly bathed in Earthshine before the new crescent moon appears.

Moonlight: The Moon is not very reflective due to its rough surface and shadows from its various features, which gives rise to a soft, fairly monochrome lighting effect on Earth's landscape, often described as silvery. The light reflected from Earth back from the Moon is not bright enough to activate the colour receptors in human eyes. These light-sensitive rods react to bright light in order for us to see in colour. The light from the Sun is half a million times brighter than that from the Moon and able to activate the colour sensors in our eyes. (Grego, 2003).

Mares: The darker areas (mares) are lava-filled plains, and you can make out the shapes quite easily. The ones to the right-hand side are Serenitatis and Tranquillitatis (the latter being the site for the 1969 Moon landing). They are named seas (Latin *mare*) because, to early observers, these darker areas looked like lunar oceans.

Craters: These are formed as a result of objects such as asteroids or meteorites hitting the Moon. The rock is forced out upon impact pushing the rock sideways and upwards forming a crater wall, with the material falling back down and forming rays emanating from the crater. Central peaks are formed when the bedrock rebounds or bounces back following impact. You can have fun demonstrating this by dropping a marble into a tray of flour covered with a thin layer of cocoa.

Rays: The bright lines can be seen clearly under a Full Moon as they radiate out from craters such as the magnificent Tycho. The lighter the rays the younger the crater as the lighter material is thrown out on impact onto the older darker material. This effect is nicely demonstrated in the crater making activity above.

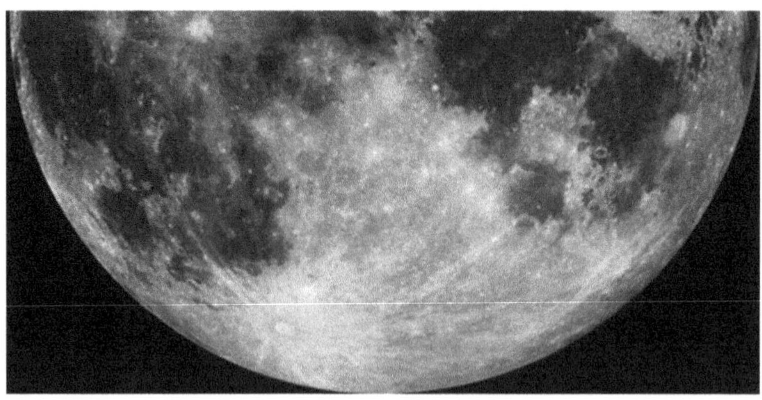

*****

Some ideas to further enjoy observing the Moon

A great way to start learning about the features of the Moon in more detail, is to look out for particular features visible during the different phases using binoculars, keeping a notebook and pencil handy for sketching. The act of sketching will help you see and understand the features. You could take a look at a list of craters and explore one or two that are of interest to you.

There are certain conditions which make viewing features easier. The boundary line which crosses the Moon bringing either darkness or light in its wake is called the terminator. As this line moves across its surface, different features are highlighted, as the low angle of the Sun casts shadows or brings features into sharp relief.

A Moon atlas or map is useful for getting to know which features will be visible during which phase. A Full Moon is difficult as the reflected light is often too bright to make many useful observations other than locating the major mares and craters but makes plotting its position in the sky a bit easier.

Some features visible during particular phases

Waxing Crescent

The round **Mare Crisium** and the larger one, **Fecunditatis**, below it are clearly visible during this phase. These are on the right-hand side of the Moon when viewed through binoculars or with the naked eye.

Craters **Lockyer** and nearby **Janssen** (which are also visible at First Quarter) are of interest here because of their names. As a member of the Norman Lockyer Observatory in Devon, I have a particular interest in these people. The craters are named in honour of Norman Lockyer, a British civil servant/astronomer and all-round fascinating character, and Pierre Janssen, a French astrophysicist who, independently of each other, developed a new technique for analysing the chemical composition of the Sun. In honour of their achievements the French government awarded a joint medal to them. A replica is on display at the Observatory in Sidmouth.

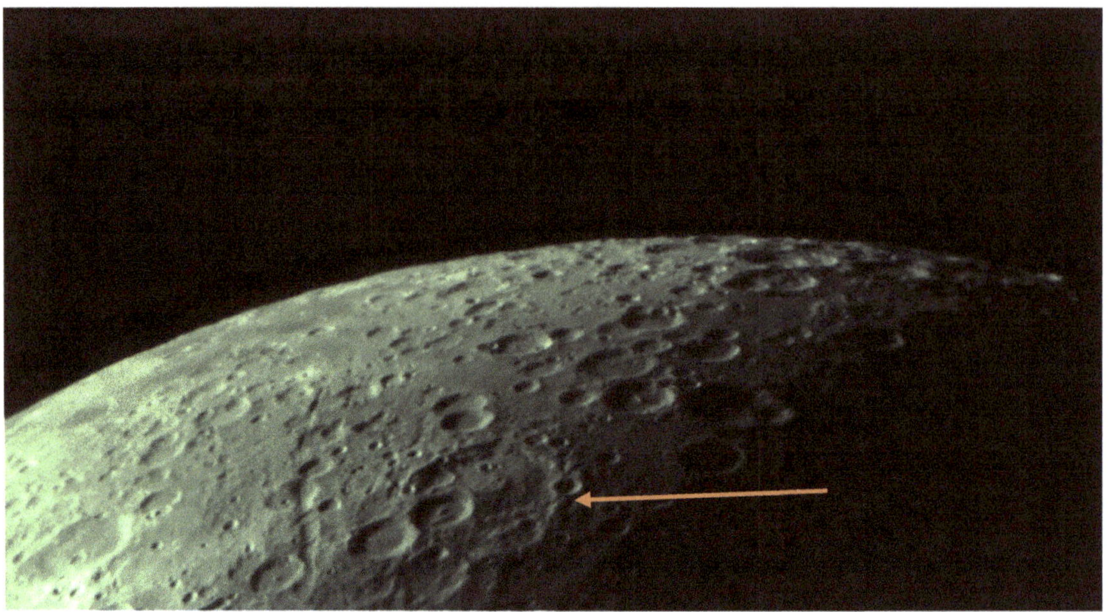

Lockyer and Janssen Craters David Strange NLO

First Quarter Moon

Mare Serenitatis is a large lava-filled plain just above the more famous mare, **Tranquillitatis**. This latter is obviously a must-see mare! It contains the landing site of the first lunar landing amongst others. It is strewn with large rocks which proved challenging to Neil Armstrong, who had to do some last-minute adjustments as the boulders were somewhat larger than originally thought. Astronaut footprints remain due to the lack of wind to erode them. Mare Tranquillitatis is 872 km/542 miles in diameter – plenty of room for landing!

Inside Mare Serenitatis (diameter 674 km/419 miles) is the **Serpentine Ridge** which is 563 km/350 miles long and width at its max is 11 km/7 miles, with the central peak rising to 304m/1000 ft. It is thought to be the best-known example of a wrinkle ridge (bendy), having been created by the cooling and contraction of basaltic lava.

The Lockyer Crater is also visible (close to Janssen).

Waxing Gibbous

Copernicus is one of the most readily identified craters by its sheer size and location, close to the lunar equator and a bit to the west. It is a great one for practising sketching skills. An artist friend, John Meacham, made this sketch:

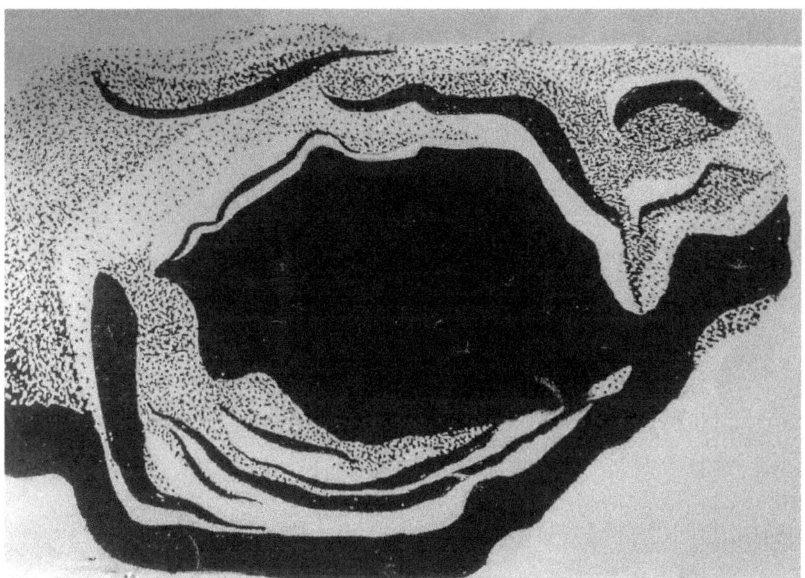

The crater is 92 km/57 miles in diameter with walls that are over 3 km/2 miles high with high central mountains.

Tycho is well known for its very long and spectacular ray system, the most prominent on the Moon. On its slopes rests a survey vehicle from one of the missions to survey the Moon. The magnificent ray system makes this a must-see object. Some of the rays stretch out for about 2,414 km/1,500 miles. It is in the Southern Highlands and is one of the few features easily seen during a Full Moon. This is one of the youngest craters and was probably formed during the age of the dinosaurs (at least 66 million years ago). The size of it is 84 km/52 miles across with central peaks around 3 km/2 miles high. They are particularly visible because they are relatively new. Remember, when an object hits the Moon, newer material is thrown out in rays over the older, darker surface.

Mare Imbrium has an attractive mountainous feature to enjoy. The scenic Bay of Rainbows (Sinus Iridum) is a semi-circular feature in the Apennines with 457m/15,000 feet high peaks. It is well worth trying to find. The Apennine Mountain range is to the north of the Apollo 15 landing site and is at the south-eastern part of the Imbrium basin (almost opposite the Sinus Iridum). It is particularly interesting when the Sun's rays catch the mountains, as in the sketch below (also by John Meacham).

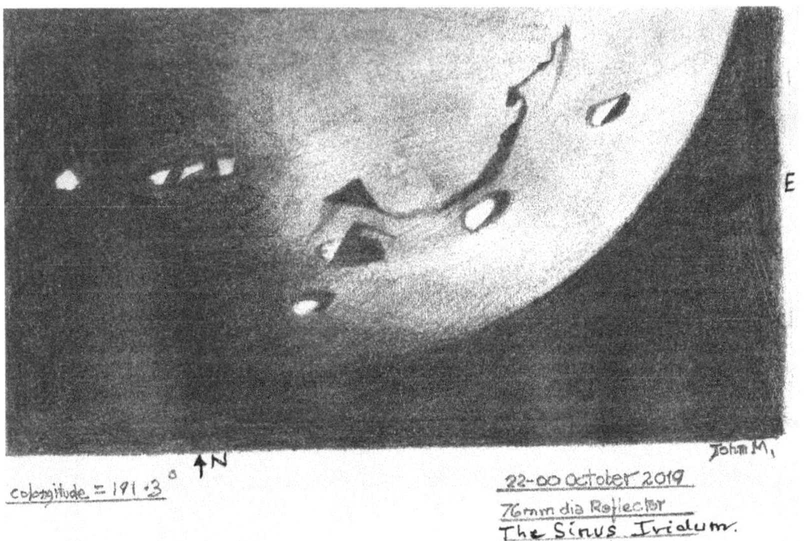

Full Moon

Tycho can be seen as a bright spot with the naked eye. The dark plains of the mares and the brighter cratered highlands show up clearly. It is a joy to look at the surface with a pair of general-purpose binoculars (cameras will need a lunar or polarising filter to cut down the brightness) to see the many features lit up in the full Sun.

On a cold, dark January evening you could have a fun session trying to locate Wolf Crater beneath a Wolf Moon. The term *Wolf* was given to the Moon appearing at this time by native Americans who heard the howling wolves as they searched for food in deep midwinter. The crater itself is in the centre of Mare Nubium, to the south-east of Mare Procellarum. It is in the middle of a fairly dark area. The outline is more like a heart than a wolf though! It is small, 26 km/16 miles wide, yet full of smaller craters with bright ejecta material to be seen spreading out from its sides. The crater is not named after the animal though but named after Max Wolf, a German astronomer, and was photographed by the crew of Apollo 16 in 1972. This crater is an example of how rewarding the detailed viewing of less well-known features may be.

NB: It is a small crater and labelled Max Wolf Crater in some books. A friend suggests finding Bullialdus Crater first, then move west to find Wolf. This crater would also be suited for seeking out in a less full phase. Happy Hunting!

Copernicus revisited. *See previous reference to this magnificent crater for full information.* At this stage of the Moon's cycle, however, you could note how it appears under the lighting effects at Full Moon compared with earlier.

Last Quarter

Plato Crater is an interesting feature visible from first quarter through to last quarter. Its dark, relatively smooth floor is made up of lava flows which have covered the central peak. The geological activity occurred over 3 billion years ago. It has been called a 'Great Black Lake' because of its smooth, dark floor.

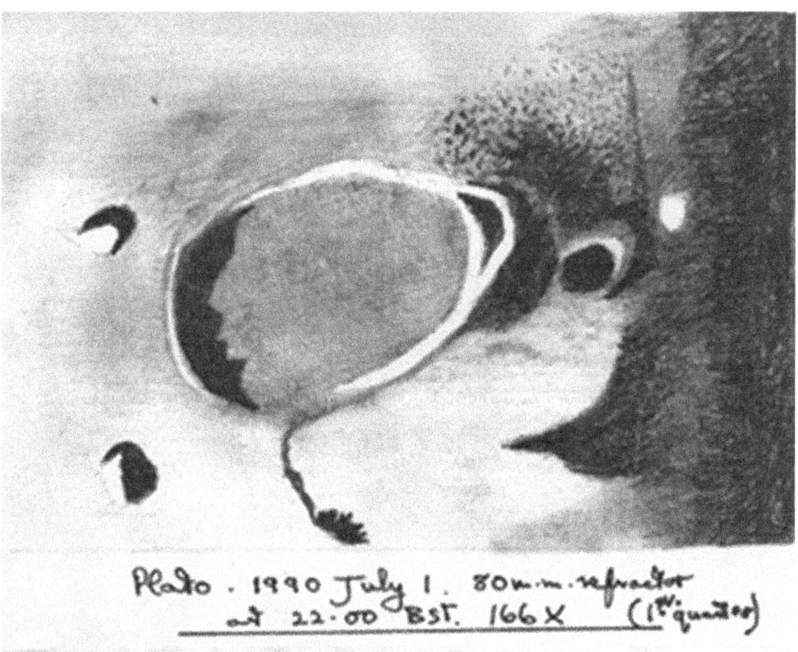

When the terminator (that division between light and dark) is about halfway across the Moon's face that we can see, a line of craters is clearly visible: **Ptolemaeus, Alphonsus,** and **Arzachel**. Initially I made a rough sketch of these and then compared them to the features on a lunar map and was able to identify them. The process of observing and sketching and identifying relatively easy targets will provide good practice in techniques for identifying other features.

Photo of Ptolemaeus chain of craters taken by David Strange, Chief Astronomer, Norman Lockyer Observatory, Sidmouth

Looking at these particular three in more detail: the features of Ptolemaeus are best viewed when the Sun is at a low angle, as during Last Quarter. It has no central peak and is 64 km/40 miles wide. The middle of the chain of craters is Alphonsus which slightly overlaps that of Ptolemaeus. It contains a few smaller craters and has a steep central peak. Finally, there is Arzachel with a terrace area to look out for. It shows clearly as a circular crater

Also clearly visible is Mare Imbrium together with the Bay of Rainbows feature on the rim of the Jura mountains opposite the Apennines. The 'mountain ranges' which encircle this mare are actually crater walls. Mare Imbrium is a vast crater of 250 km/155 miles diameter the walls of the Apennines rising to 457 m/15,000 feet.

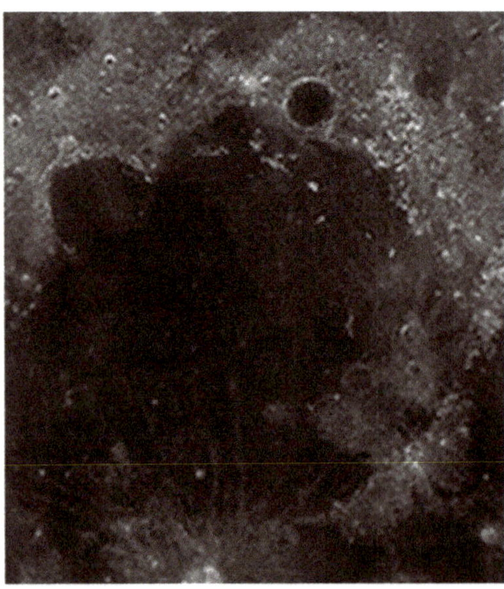

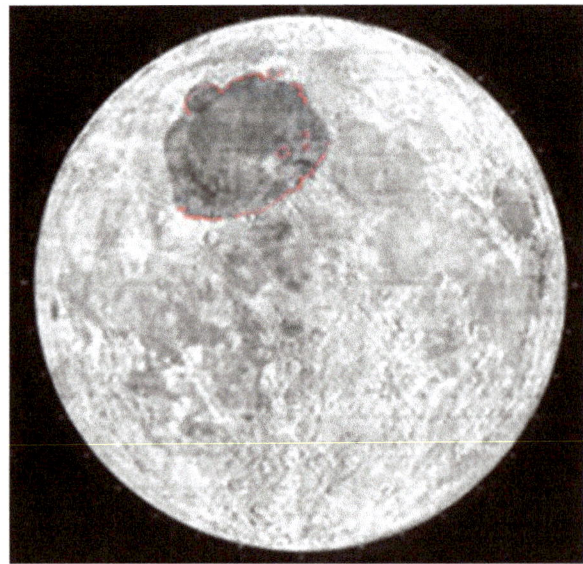

Waning Crescent

This is the Moon between 14-24 days in its cycle. Gassendi Crater (on the northern edge of Mare Humorum) contains a central peak and a tremendous system of rilles (long, narrow valleys with steep sides). This crater is 89 km/55 miles in diameter. The wall has been badly damaged in places by lava flow material.

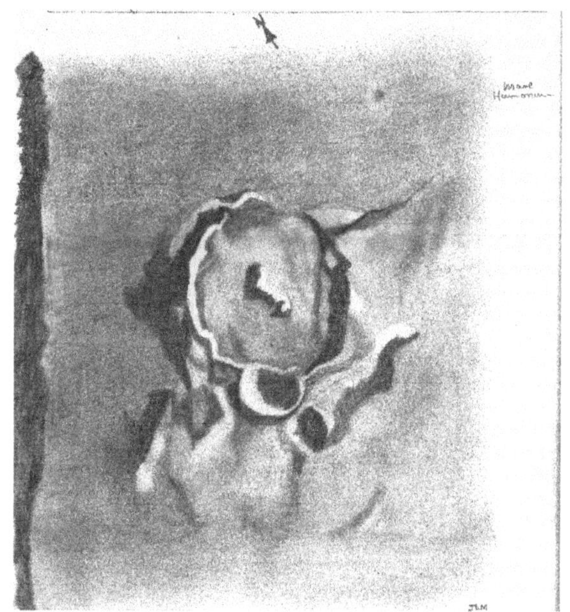

A sketch of Gassendi Crater made by John Meacham

You don't need to own a telescope, just visit a local observatory, or have a day out to one. Or perhaps build your holiday around one! And do get in touch with your local astronomical society.

*****

Some additional information about the Moon

Glimpses of the *far side*

The far side of the Moon is a mysterious place holding endless fascination for humans. When astronaut Michael Collins was left in charge of the orbiting spacecraft during the first Moon landing in 1969, he commented that he felt alone but not lonely as the spacecraft flew around the far side of the Moon, making him the furthest human in space at that time. His fascinating story can be read in his book, *"Carrying the Fire"*. It is wrong to call the unseen side the Dark Side because it is not always dark; it is simply not visible to viewers from Earth. We can catch tantalising glimpses of this heavily cratered far side though due to the effects of libration (the 'wobbling' of the Moon). The Moon protects us from all sorts of potentially hazardous meteorites over time, as they hit the Moon instead of Earth.

> *Libration* *enables us to see some features of the far side, and, over time, we can see 59% of the Moon's surface. The Moon rocks a bit east and a bit west (libration in latitude) and north and south (libration in longitude, moving up and down relative to Earth giving us tantalising glimpses further north and south). In the words of the late, great Sir Patrick Moore, the Moon nods and shakes its head!*

A good way to see libration in action for yourself is to look at two well-known features, such as Mare Crisium and Plato Crater, and notice the way the shapes change over a period of 29½ days. This timeframe is called a lunation, which is the time between one new moon and the next. Mare Crisium in the northwest seems more oval at times and sometimes more circular.

Just a bit below the north pole is the dark shape of the crater Plato. Again, this looks more circular at times and then looks more oval at another time. Small binoculars are fine for observing these features.

If you have a look at Mare Crisium shortly after New Moon (waxing time), you will be able to glimpse Mare Marginis (the Marginal Sea!) for a while and in the waning half of the cycle, craters Grimaldi and Riccioli appear to move from the western limb of the Moon towards the centre which will show you libration in action.

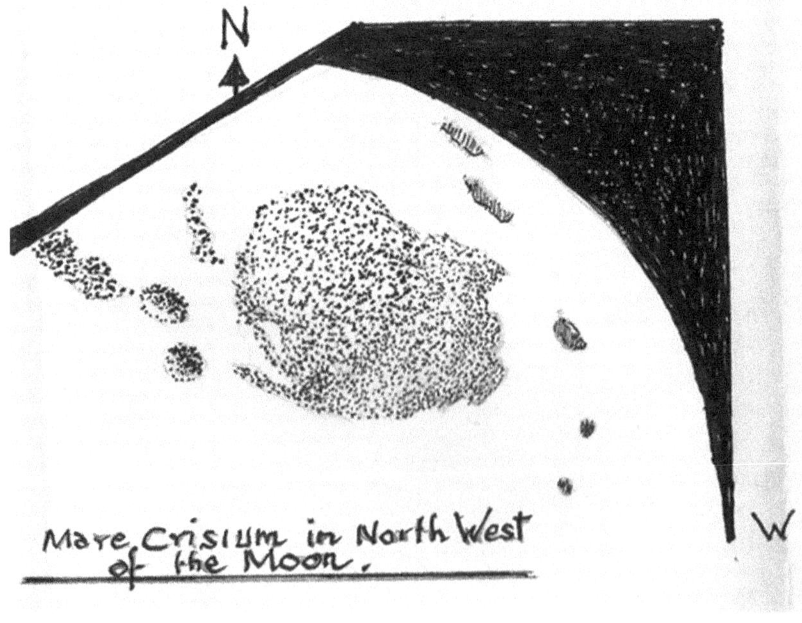

Mare Crisium in North West of the Moon.

By a fortuitous stroke of good fortune, the Moon came to my local cathedral where I was able to take a photograph of the elusive *far side*!

The internally lit sculpture made by Luke Jerram, was suspended from the cathedral's famous nave ceiling,

features high-resolution NASA imagery of the lunar surface. At an approximate scale of 1:500,000, each centimetre of the artwork represents 5 km of the actual moon's surface.

Orbit and Rotation

On the face of it, these would seem fairly straightforward. The Moon goes around the Earth while both go around the Sun. However, our view of the Moon from Earth makes it more complicated due to the tilting of the Earth; the Earth has a 23-degree angle on its axis. This means it does not go around the Sun in an upright manner. The Moon orbits the Earth at a 5-degree angle to the Earth's orbit of the Sun, with both then going around the Sun. This is best illustrated in a diagram (see below).

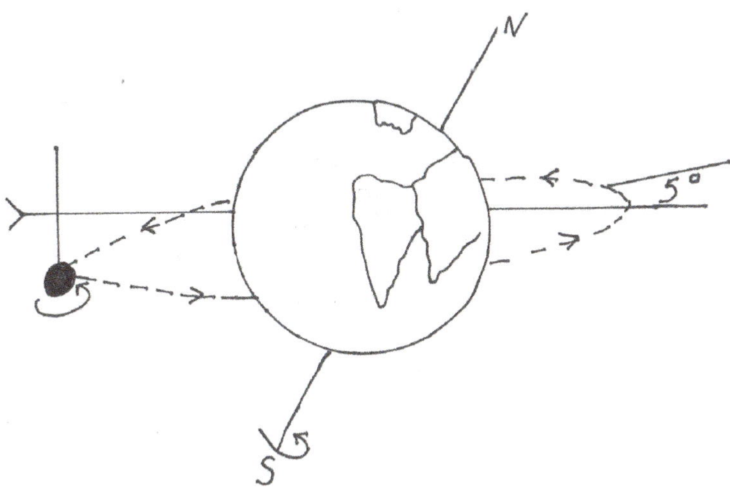

Sometimes you will see the Moon high in the sky on one occasion and right down on the horizon the next, because both the Earth and the Moon have elliptical orbital paths with the Moon on a slightly different plane from that of the Earth. What happens in midwinter is that the Sun is at its lowest while the Full Moon is high in the sky in the early morning; in midsummer, the Sun is high in the sky with the Moon at the lowest point of its orbit. Thus, in the autumn and winter months the Moon appears as a 'Supermoon' when it is low on the horizon. This term is evocative and much loved by the press. However, it helps boost interest in astronomy as do other terms, such as Blood, Harvest, Blue, Flower, and Strawberry, given to the Moon at different times of year, yet it remains the same Moon. The variance in its size is due to the perspective of a viewer from Earth. You can prove this by holding your thumb to the Moon and see that it covers it whether it looks large or small.

The Moon rises in the east and sets in the west but day by day we see it move across the sky from west to east as the Earth spins underneath the Moon while it is slowly orbiting and, even more slowly, rotating. The Moon moves from west to east by one Moon diameter per day. The total time taken for the Moon to rotate once upon its axis is the same time that it takes to orbit the Earth, a system known as tidal locking. The result is that only one side or face of the Moon can be seen from Earth. If it were not rotating, we would see the back part of it come into view during every orbit.

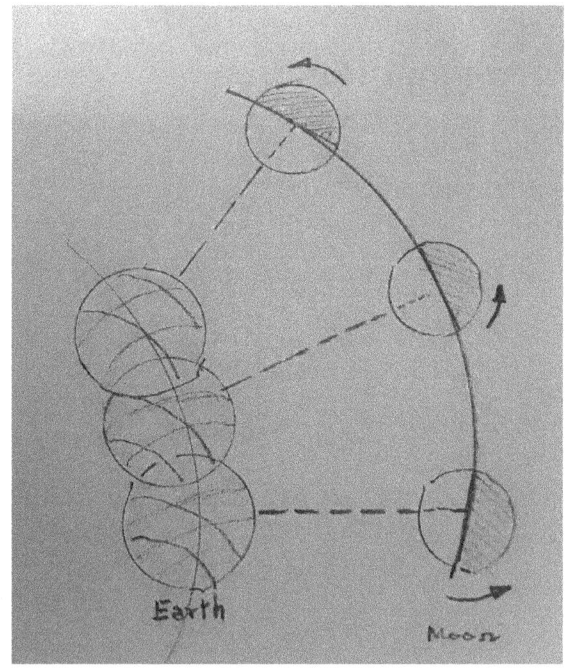

The Moon has a two-week long day compared to Earth's 24 hours, with temperatures ranging from 150°C by day down to **-270°C** at night. When two Full Moons occur in one calendar month, these are called Blue Moons (the term originating from a 16th century expression for a rare event). The colours of the

Moon are due to dust in the atmosphere. Red is the more likely colour, for example due to dust from volcanic eruptions, although in 1883 people reported seeing a blue-coloured Moon following the eruption of Krakatoa, resulting from the scattering of light on tiny ash particles. Two Full Moons in one month occur because, while the Moon makes an orbit around the Earth every 27.3 days, the period of the Moon's phases range over 29.5 days due to the movement of the Earth around the Sun. Therefore, in any month with more than 28 days, two Full Moons can occur. A look at an orrery either in real life in a science centre, at an observatory, or a virtual one on the internet will help explain this further. An orrery is a mechanical model of the solar system.

At its furthest point, or apogee, the Moon is nearly 406,000 km/252,000 miles away and at its closest, or perigee – when the Moon looks like a Supermoon – around 360,000 km/223,000 miles away. The average distance is 400,000 km/ 250,000 miles away. Although the Sun, Moon and Earth are aligned, because the Moon's orbit is not exactly in the same plane as Earth's orbit around the Sun, they rarely form a perfect line. When they do, we have a lunar eclipse when Earth's shadow crosses the Moon's face.

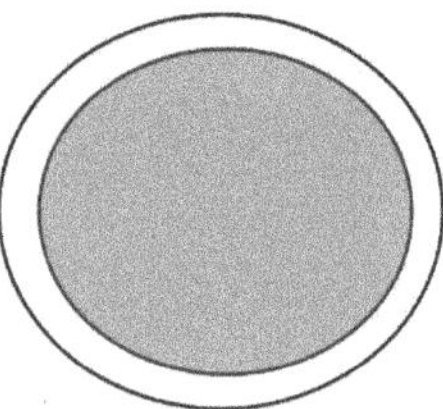

SIZE OF MOON AT CLOSEST (PERIGEE) TO EARTH
AND AT FARTHEST (APOGEE) FROM EARTH

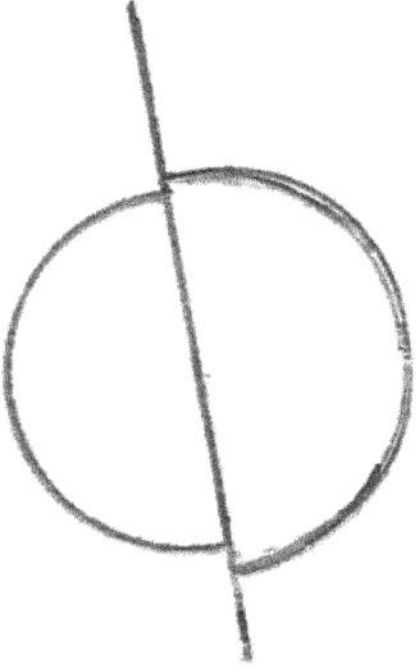

*A partial eclipse of the Moon as seen by the author on
7 September 2006 at 20:15 BST from Roundball Hill, Honiton, Devon.
If you look closely, you can see the top is 'sliced off',
where the Earth is blocking the light to the Moon.*

*And in 2024 at a less convenient time, around 03:15, when there was
another partial eclipse.*

A lunar eclipse occurs when the Earth comes between the Moon and the Sun. This is nicely illustrated in the above photographs. The keen-eyed amongst you will also notice there is a gap of about 18 years between the photographs. Other lunar eclipses have occurred in between the two examples but the Moon also has an 18.6-year cycle when the Moon reaches the extreme highest or lowest points on its orbital tilt before tilting back again. At the end of each 18.6-year cycle, the Moon appears to standstill – a 'lunarstice', much like a solstice when the Sun appears to stand still. This cycle is the range of movement within the Moon's orbital tilt. 2006 and 2024 are examples of each end of this cycle.

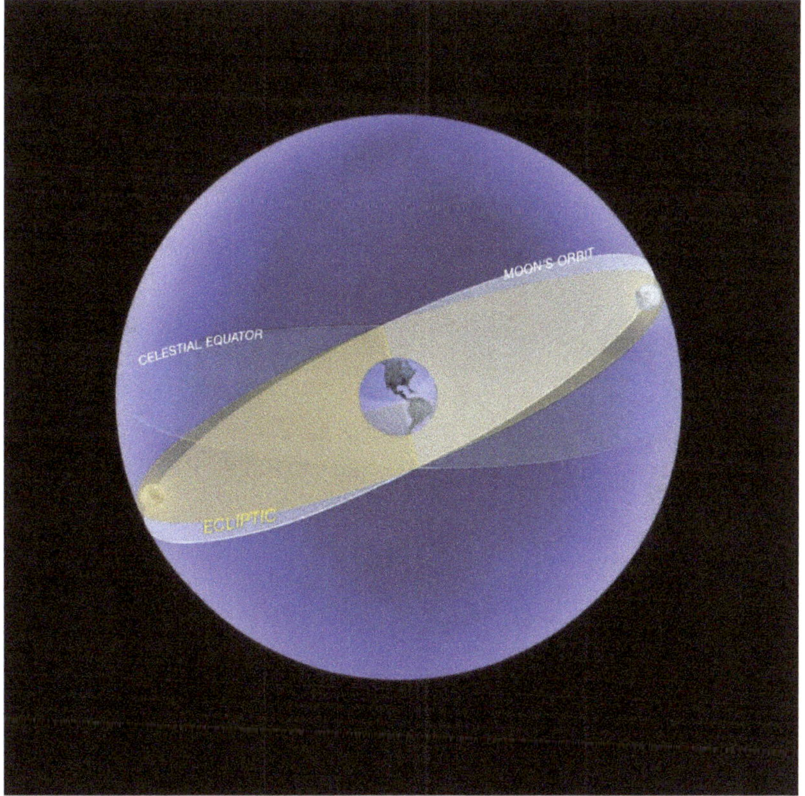

Illustration by Griffiths Observatory showing the range of movement of the Moon.

When the Moon is between the Sun and the Earth, we have solar eclipses. These are events to be enjoyed now as total solar eclipses may not occur in the future due to the Moon moving further away from the Earth at a rate of 3.8 cm/ 1.5 inches a year. At present, the distance and relative sizes give rise to the quirk that, as viewed from Earth, the Moon covers the entire Sun. Why does this work? The Sun is 400 times the diameter of the Moon and is 400 times away. It is simply a matter of perspective and current distances.

It is in relation to water that the Moon has the greatest effect on Earth. High and low tides on Earth are caused by the Moon pulling the water as it orbits, which causes it to bulge thus creating tides. If the Moon was not there, there would be less tidal activity. The gravitational pull of the Sun on water is weak due to the distance the Sun is away from Earth. However, with the addition of the Moon's gravitational pull, the tides range from very low to very high according to where the Moon is in relation to the Sun. When both are pulling from the same direction, i.e. in spring and autumn, there are high 'spring' tides which cause much damage from erosion and movement of sand.

Is there water on the Moon?

Yes. In 2009 The Lunar Reconnaissance Orbiter (LRO) was launched by NASA to make a comprehensive atlas of the Moon's features and resources. This mission found evidence of water ice at the south pole as well as finding evidence of recent geological activity on the Moon. Newly formed craters from recent meteorite impacts were imaged. The mission also tested space borne laser communication technology as well as the data gathering.

New data from the LRO was analysed in 2013 which further pinpointed areas near the south pole where water is likely to exist. Research found that hydrogen, the main ingredient in water, is trapped within the lunar soil. There was increasing evidence to suggest the presence of frozen water in the hydrogen-rich areas near this south pole area. Data from a 2018 Indian mission, which was recently re-evaluated, also confirmed the existence of water around the lunar poles.

As the Moon is only slightly tilted on its axis it means that some parts of the craters are permanently in shadow giving rise to very cold temperatures and enabling ice to remain. These craters are among the coldest in the solar system, comparable to those on Pluto. The temperatures stay below -150°C.

Human life could not live unsupported on the Moon though due to the lack of atmosphere. There is not enough oxygen to sustain life as we know it.

Moon Landings

In 2019 NASA and the World celebrated 50 years since the first Moon landing on 29 July 1969. For me, it still seems like yesterday when I sat watching the repeat of the Moon Landing on Christmas Day 1969. I couldn't get enough of it. I had followed the mission avidly in the July when the whole world held its collective breath as the astronauts returned to Earth. And I sat, equally entranced, during that Christmas Day repeat. The Moon has captivated me ever since. It is hard to comprehend that only 12 men (no women as yet) have walked on the Moon. And sobering to think the footprints they left will probably outlive the human-race itself. However, in all probability, humans will revisit the Moon one day. The astronauts on the International Space Station are busy with ongoing experiments for the benefit of future space travel and landings.

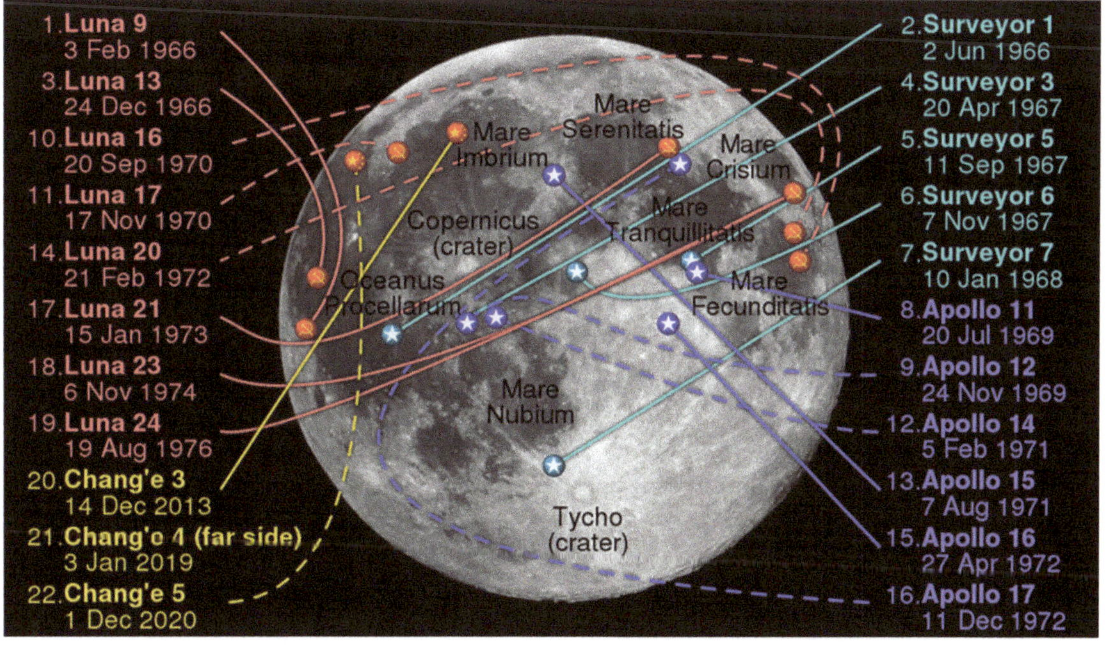

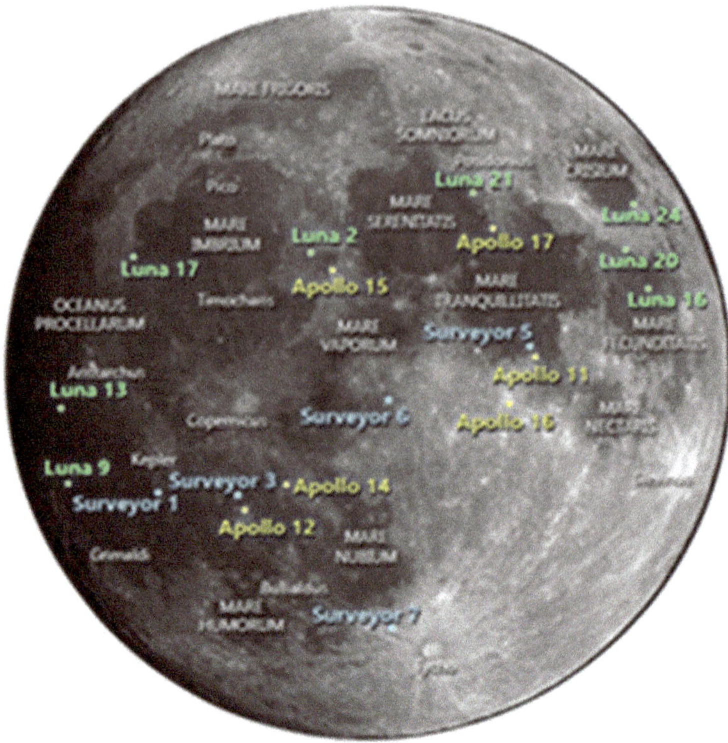

Site of Apollo 17 landing site: Taurus-Littrow Valley 11 December 1972
Picture credit: John Meacham

Bouncing

If you ever get the chance to stand on the Moon, you'll see that the effects of gravity are much reduced. The gravity on the Moon is only 16.5% that of the Earth's. In other words, if you weigh 100kg on Earth, you will only weigh 16.5kg on the Moon. This is why astronauts are able to bounce on the Moon.

The atmosphere is very, very thin, so the footprints remain because there is no wind or rain to erode them quickly. Those footprints will be there for many more millions of years. However, they are slowly being eroded by tiny meteorites, and general displacement of material. Gravity is very weak on the Moon, which means lighter elements, such as oxygen, escape, making life for humans impossible. There is much similarity between space travel and deep-sea diving with parts of astronaut training carried out in deep water. Thankfully on Earth, gravity is so strong that it retains all the elements necessary for breathing and life!

The thin atmosphere has quite an effect on the temperatures on the Moon. If you touched the surface with your bare hands it would feel like touching the ground here – warm in the sunshine and cold in the shade. It is unlike Earth as the ground does not then heat the air above it.

A cold, lifeless world but vital for the richness of life on Earth, the Moon is an endless source of wonder.

Zooming in on Earth reveals a slight bulge on the right side: that is the Moon, just peeking out from behind Earth. At the time the image was taken, Parker Solar Probe was about 43 million km/27 million miles from Earth.

PART 2

A GUIDE TO SOME MAJOR NORTHERN
HEMISPHERE CONSTELLATIONS
FROM SEPTEMBER TO AUGUST

A GUIDE TO SOME MAJOR NORTHERN HEMISPHERE CONSTELLATIONS FROM SEPTEMBER TO AUGUST

Why begin this section in September? Well, the evenings are becoming darker earlier, and it is still quite warm... a lot warmer than the clear, cold nights of February when the atmosphere is still which, although making for better observing, one can easily become baffled by the number of stars visible. So, by starting to learn the stars from September means that by February you will have gradually learned some stars in order to better find your way around the night sky... and become more acclimatised to being outdoors.

The stars forming constellations are named according to the Greek alphabet, with Alpha being the brightest. Thus, the brightest star in Ursa Major will be *alpha Ursa Majoris.*

The selection described in this section were chosen for their accessibility and interest, broadly in the order from autumn through to summer. Some of the constellations are visible all year round, the circumpolar ones, those encircling the Pole Star. As there are a total of 88 constellations, it is best to learn the main ones visible (this book is aimed at readers in the Northern Hemisphere) and those more widely known. It was Ptolemy who collated the constellations identified and named by ancient Greeks. He produced an astronomical manual, The Almagest, in the second century AD.

THE PLOUGH
WITHIN URSA MAJOR

This image to the left, taken by Akira Fujii with a backyard telescope, shows the Location of the Hubble observations near the Big Dipper. Credit: A. Fujii

The unmistakeable image of the Plough (also known as the Big Dipper) is a good place to begin to find one's way around the night sky as it is a circumpolar constellation and thus visible all year round. It is an asterism within the large constellation known as Ursa Major; that is to say the Plough is a pattern of stars within a larger constellation. In ancient times, the Bedouin Arabs referred to the asterism as a funeral cart while Native American Indians saw it as a bear with warriors hunting it. In Roman times it was a bear representing the story of the jealously felt by Jupiter's wife Juno for Jupiter and his love for Callisto. Callisto's son became Ursa Minor as Juno changed Callisto into a bear in punishment.

Pointers to the Pole Star
The northernmost star is Dubhe, a double star, and with Merak point towards the Pole Star. They are the two stars in the bowl furthest from the blade or handle. From Merak continue to Dubhe and beyond until a bright star is reached: that is Polaris or Pole Star.

Polaris the Pole Star (or alpha Ursae Minoris)

Polaris is about 400 light years away. It is the main star in the constellation of the Little Bear or Ursa Minor and for centuries has been used by sailors to determine their latitude according to its height above the Northern horizon. The altitude of Polaris is the same as your latitude. The Pole Star stays (at least it does change but imperceptibly!) as a beacon over the North Pole which is why it is a crucial navigational aid. The diagram here illustrates its location:

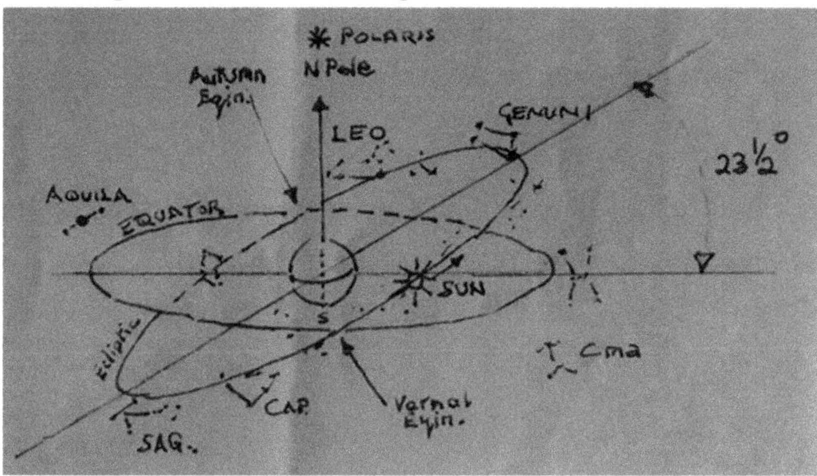

Mizar

Mizar has a close companion star, Alcor. They are perceived as a double star but are in fact very far apart (about one light year). They were the first 'double star' to be viewed through a telescope in 1650. The two stars are part of a gravitationally bound system of 6 stars.

Alioth

It is the brightest star in the Plough and one of the brightest in the night sky. Although part of Ursa Major, the name actually means 'tail of the sheep'! It is part of the Ursa Major Moving Group. In 1869, it was discovered by British astronomer Richard Proctor that most of the stars in the Plough were found to be moving towards Sagittarius.

A variable star

Z Ursae Majoris, a pulsating red giant, is an ideal variable star for beginners. If one can observe it every 10 days or so the magnitude (brightness) will be seen to vary from 6 to 9 on the Pogson Magnitude Scale. Variable stars are grouped in four basic classes: Pulsating, Eruptive, Cataclysmic and Eclipsing Binary.

Galaxies M81 and M82

In a very short space of time, one can be drawn into the delights of deep sky objects by locating the position of some galaxies near the Plough.

M81 (Bode's Nebula) is a spiral galaxy about 12 million light years away in the constellation Ursa Major and one of the brightest in the northern sky. It is relatively easy to find in 10x50 binoculars in good dark seeing conditions.

> *M is short for Messier. A catalogue was compiled by a Charles Messier when he was comet hunting. Once an object was identified and found not to be a comet, he added it to a list to save time and effort by other astronomers.*

M81 along with M82 (Cigar Galaxy) can be used as a good demonstration of averted vision: if you have them both in the same field of view, you may see that the core of M81 becomes more apparent if you look at M82. Both galaxies are visible as 'fuzzy blobs' to the naked eye in even quite light-polluted areas.

> *Averted vision? If you look at the area of interest then look away and then look back out of the corner of your eye, you will have more light entering the lens of your eye which enables you to observe the 'fuzzy blobs' of distant galaxies.*

It is sobering to think that the Milky Way Galaxy in which our solar system resides would look as tiny to an observer from M82. M82 tends to appear as an edge on smudge through a telescope although it is five times brighter than the Milky Way.

M101

A spiral galaxy which can be seen face on. It is close to the blade of the Plough. It is 22 million light years away. It appears as a faint smudge in a small telescope and therefore offers much more of a challenge.

Some other galaxies near the Plough are M108, an edge-on spiral galaxy, M109, a barred spiral galaxy and the nearby Owl Nebula (M97), a real challenge for which you need a large telescope – or this could be something a local observatory/astronomy society may be able to show you if galaxy spotting has caught your interest.

SUMMER TRIANGLE
WITH CYGNUS, LYRA AND AQUILA

The Milky Way passing through the constellation of Cygnus

This is a recognisable triangle pattern of stars, providing guideposts to three constellations. **Vega** is in Lyra, **Deneb** (to the east of Vega) is at the 'tail' end of the constellation Cygnus, and **Altair** is the brightest star in Aquila.

Vega is the brightest of the three and is almost directly overhead in the summer; the brightness due in part to its proximity to Earth (25 light years) and also to the fact that it is 52 times as bright as the Sun. **Deneb** is a white supergiant. At 3,000 light years away, it is the most distant of the brightest stars as seen from Earth. It is about 100 times the size of our Sun. **Altair** is just 17 light years away with a fast rotation causing the middle to bulge.

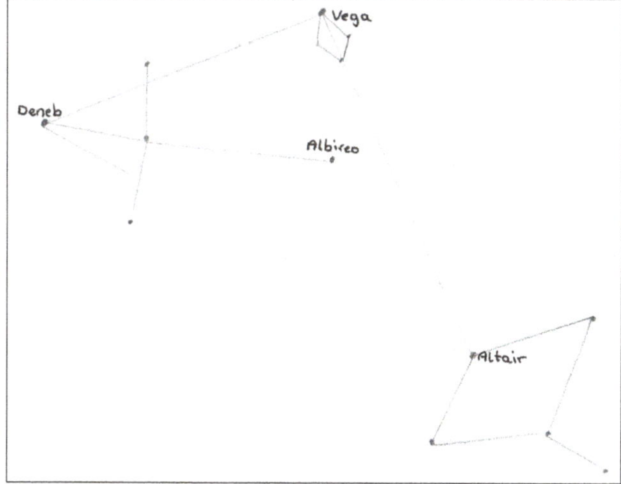

Constellation of LYRA

This was named after the *Lyre of Orpheus*.

Vega

The brightness of Vega is due in part to its proximity to Earth as well as the fact that it is 52 times as bright as the Sun. It is one of the nearest stars to us at a mere 25 light years away. Like our Sun, it is halfway through its lifespan. At one time it was the pole star and will be again due to the movement (precession) of the stars. It is a hot blue star, about 450 million years old.

Vega is an important star because it is the primary star in the magnitude scale. Magnitude measures the difference in brightness between stars. Vega has a magnitude of 0 and it is upon this star that the *apparent* magnitude scale is based; first magnitude being the brightest *as seen from Earth* down to sixth magnitude. The greater the number, the *fainter* the star. When a star is brighter than Vega, it will have a negative magnitude i.e. the bright star Sirius is -1.6. This is *apparent* magnitude, while *absolute* magnitude refers to the actual brightness of a star. This latter is worked out on a different scale related to distance, used by professional astronomers.

M57 Ring Nebula

On very clear nights, this can just be seen as a faint elliptical disk in a small telescope, although it looks stunning on long exposure photographs.

Lyrid Meteor Shower

Around 21-22 April each year can be seen this bright meteor shower which has about 10 meteors per hour. Originating in the constellation of Lyra, these meteors come from the tail of Comet Thatcher. The comet last visited the inner solar system in 1861 and is expected to return in 2276. It was first formally identified in 1861 by an A. E. Thatcher. It has a very large orbital path around the Sun. With records of it going back for 2,700 years, this is among the oldest known meteor shower.

Constellation of CYGNUS

Cygnus is Latin for The Swan

Deneb

Deneb means tail and is indeed situated at the rear end of the constellation. At 3,000 light years away, this white supergiant is the most distant of the brightest stars as seen from Earth. It is about 100 times the size of the Sun. This star is the starting point for finding Sagittarius and thence Scorpius, to the west. The cross points northwards towards Cepheus and southwards to Sagittarius.

Albireo

Albireo is at the front or head of the Swan. It is a yellow/blue double star, a pair of stars that appear close together as seen from Earth. One of these stars is a third magnitude orange giant which has exhausted its supply of hydrogen, together with a fifth magnitude blue/green star. The blue/green colour indicates it is much younger than its partner. It is a very pretty object to view through binoculars and telescopes and is justly popular. I remember my own excitement and pride when showing someone this star through binoculars; it really does look like two stars, and I was proud that I could easily find it.

Some other features to look for around Cygnus are:

M39 Open Cluster, which is actively producing new stars replacing those that move away from the cluster.

NGC7000 This nebula is visible through binoculars, located just to the north of Deneb. It is also known as the North American Nebula, as it looks like that continent.

NGC6826 Called the Blinking Planetary, this appears as a pale blue disk. Although it is as bright as a 9th magnitude star, the light is spread out over a disk instead of being concentrated at a point. The nebula is visible with averted vision. It is 2,000 light years from us. It appears to blink because when you stare at it, it seems to disappear – giving a 'here again, gone again' effect. Planetary nebulae are what remain when a red giant explodes and dies; so-called because they looked like planets to early astronomers. Clouds of gas are thrown off during the explosion and remain as planetary nebulae.

NGC6992 Veil Nebula – mainly visible only on long exposure photographs. It is the remnants of a supernova that exploded 5,000 years ago. In a really dark sky, the western part is just visible with averted vision.

M29 Open Cluster of about 50 stars, 5,000 light years away. It is located just below the star Sadr.

M56 Globular Cluster. At 13 billion years old, it is three times the age of our Solar System, dating back to the beginnings of the Universe. The 100,000 stars are in a cluster 10 light years in diameter and 40,000 light years away from us. These clusters age and die as a group since the stars are formed at roughly the same time and do not generate new stars, in the way Open Clusters do.

The Milky Way passes through Cygnus and is clearly visible on occasions. The band of stars is part of one of the spiral arms of our own galaxy. On a dark night it is visible as a background pathway of dim stars. Within the Milky Way lies the dark lane of dust dividing the bands of stars, which is also visible during darker nights.

Constellation of AQUILA

This is Arabic for Flying Eagle.

Altair

This is one of the nearest to us at just 17 light years away. It is a fast-rotating white star, rotating in 9 hours. This rapid rotation causes the middle of the star to bulge. Just above Altair can be seen the small arrow-shape of Sagitta, one of the smallest constellations. Within Sagitta is the globular cluster of M71 which shows as an elongated misty patch through binoculars or telescopes.

M27 The Dumbbell Nebula is a more challenging observation target and another one for which to visit an observatory, unless you have access to a telescope.

SAGITTARIUS

Credit: ESA, NASA & Akira Fujii

Containing the so-called Teapot asterism, this is an exciting constellation generally heralding the start of the serious observing season… or a sad one indicating the end of summer. A lot, as always, depends on one's perspective, location, and interest. This constellation is fairly low on the horizon with its distinctive 'teapot' shape visible. Sagittarius is Greek for 'archer'. He is a centaur with an arrow pointed at the heart of the constellation of Scorpius, which is represented by the bright red star Antares. There are many deep sky wonders within this constellation which you may like to have a look for. Initially, become familiar with the teapot shape that is just above the horizon.

M20 Triffid Nebula

M20 is a popular object for amateur astronomers because it is quite bright and easily observed in small telescopes. It has a diameter of 42 light years and lies at a distance of 5200 light years from Earth. The nebula's name means 'divided into three' and refers to the object consisting of three types of nebulae: an emission, a reflection, and a dark nebula. There is also an open star cluster which forms M20 along with the nebulae.

M22

One of the finest globular clusters in the sky, visible through binoculars and by the naked eye. At just 10,000 light years away, it is one of the closest to us. It is elliptical in outline.

M23

Open Cluster of 150 stars, visible through binoculars.

M24

This is a rich star field visible by naked eye. Binoculars show it as a field of stardust.

M25

A cluster of about 50 stars, visible through binoculars, within which lies a yellow supergiant.

Milky Way

The spiral arm of our galaxy passes through Sagittarius and contains rich star fields, numerous clusters, and nebulae, which make for further rewarding exploration of this area.

M8 Lagoon Nebula

This beautiful cosmic cloud is a popular stop on telescopic tours of the constellation Sagittarius.

The Lagoon Nebula is an active stellar nursery about 5,000 light years away, in the direction of the centre of our Milky Way Galaxy. It is similar in size to the Orion Nebula, about 30 light years in diameter. It is divided by a dark rift which makes it appear like a lagoon, hence its name.

Some remarkable features can be seen in this sharp picture above, showing off the Lagoon's filaments of glowing gas and dark dust clouds. Twisting near the centre of the Lagoon, the bright hourglass shape is the turbulent result of extreme stellar winds and intense starlight. This view is a colour composite of images captured while M8 was high in dark, Chilean (NASA) skies. The rift was discovered independently by John Flamsteed in 1680, Guillaume Le Gentil in 1747 and Charles Messier in 1764. It is great to note that Le Gentil had a positive astronomical experience after his somewhat unsuccessful chasing Venus exploits… (another story for another time.)

M17 Swan or Horseshoe Nebula

This appears as an elongated smudge with a sprinkling of stars either side. It is a cold, dark cloud of gas and dust, silhouetted against the bright nebula of IC434. The bright area at the top left edge is a young star still embedded in a stellar nursery of gas and dust.

CASSIOPEIA

This huge 'W' in the sky is an excellent guidepost to the Andromeda Galaxy which can be observed with the naked eye on a fairly dark night.

M31 Andromeda Galaxy

To view this using just your eyes you need to get the hang of averted vision, which involves looking at the object out of the corner of your eye while trying to view the object at the same time. This allows more light to enter your retina thus providing you with a view of the galaxy. Andromeda Galaxy appears as a fuzzy blob below the right-hand 'V' of the 'W' of Cassiopeia. Viewing through binoculars may allow you to see the central bulge, while through a telescope you may glimpse its spiral arms. The nearest galaxy to our own, it is about 2.5 million light years away from us and around 100,000 light years across. It has a satellite galaxy M32 which is similar in diameter.

Alpha Cassiopeia and Beta Cassiopeia

These are ideal for becoming familiar with magnitude. The difference between is small (2.5 and 2 respectively) but you can check for differences by looking up at them and rolling your eyes around the area where they are on a dark evening.

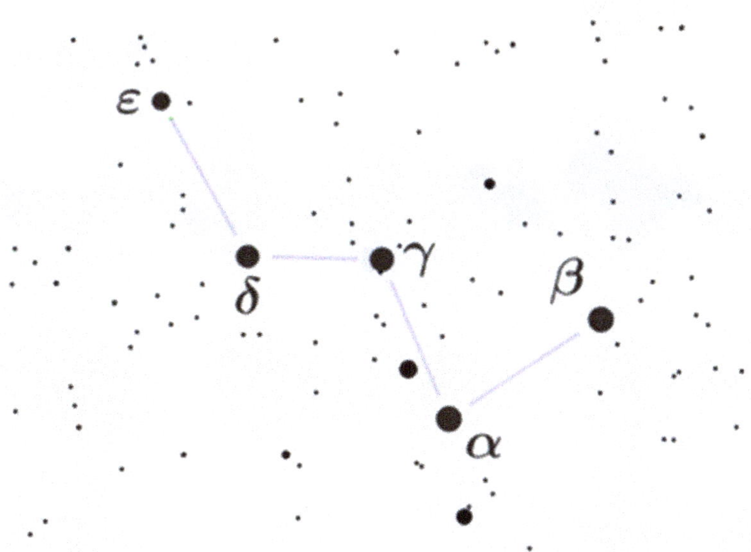

M52

Binoculars will show you this wonderful open cluster, which contains about 100 stars.

Gamma Cassiopeia - Variable Star

This is a variable star which is very unstable. It is a giant star likely to change brightness from 1.6 magnitude to 3 magnitude caused by its varying diameter. It is worth keeping an eye on. It is about 650 light years away.

ANDROMEDA

Credit: A. E. Kingston

This constellation represents a further opportunity for locating the Andromeda Galaxy. I personally find it easier to locate using the constellation of Cassiopeia, hence also the inclusion of the galaxy in that earlier section.

M31 Andromeda Galaxy

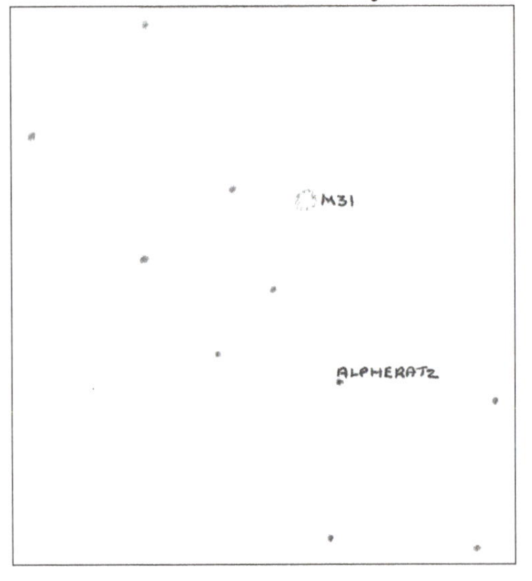

To find it via the constellation which bears its name, you need to locate the star Mirach, which is halfway between Alpheratz and the topmost star of the northern line of Andromeda, Almaak, and look to the south of the southern faint line of stars. Andromeda Galaxy is the top point of a triangle between it, Alpheratz and Almaak. To view the galaxy using just your eyes you need to get the hang of averted vision, which involves looking at the object out of the corner of your eye and trying to view the object at the same time. This allows more light to enter your retina thus providing you with a view of the galaxy. Andromeda Galaxy appears as a fuzzy blob. Viewing through binoculars may allow you to see the central bulge, while through a telescope you may glimpse its spiral arms. The nearest galaxy to our own, it is about 2.5 million light years away from us and around 100,000 light years across. It has a satellite galaxy M32 which is similar in diameter.

Alpheratz

This star at the tip of the constellation is shared with the Great Square of Pegasus (the body of the horse in the constellation of Pegasus consisting of four stars: Alpheratz plus Markab, Scheat, and Algenib). By locating this bright star, you should be able to see the line of stars making up the northern side of Andromeda constellation.

Almaak

This is a beautiful blue/yellow double star.

Square of Pegasus

This is an asterism forming the body of the horse in the constellation of Pegasus (the winged horse in Greek mythology), located below Cassiopeia, and to the left of Cygnus. The square comprises four stars: Alpheratz, a blue-white star at the head of the Andromeda constellation, with the other three within Pegasus: Algenib which is 9 times the mass of the Sun, Scheat, an old red giant and Markab, another old giant star. These four stars frame a fairly empty square of sky, making them relatively straightforward to locate.

WINTER HEXAGON

Hexagon equals 6... or does it? This is a fun asterism rather than a constellation. The 6 bright stars of the pattern are visible early on a winter's evening. Once you have located the main 6, it makes it easier to find other constellations and objects.

1. Sirius

Greek for 'scorching' or 'sparkling' due to its extra-twinkly nature. This white star is hot with a surface temperature of 9,940°K. It is known as the Dog Star because it is the brightest star in the constellation Canis Major (the greater dog), which is one of the two dogs belonging to Orion. The other dog is nearby, Canis Minor with Procyon the lead star. At a distance of only 8.7 light years away (or 80 million million kilometres or 50 million million miles), Sirius is one of the closest stars to the Sun. It is the brightest star in the sky with a companion star, Sirius B, which orbits it every 50 years. This companion star is bright in its own right but somewhat overshadowed by its neighbour.

2. Procyon

In Canis Minor. Appropriately, it forms a triangle with Betelgeuse and Sirius; as the two dogs, Canis Major and Canis Minor are the hunting dogs belonging to Orion the Hunter. This triangle is an asterism within an asterism! At 11.4 light years away, Procyon is also one of the nearest to our Sun.

3. Pollux

This is a huge star, nine times the diameter of our Sun. It is an Orange Giant, which means it is an older star nearing the end of its lifespan. It is 34 light years away.

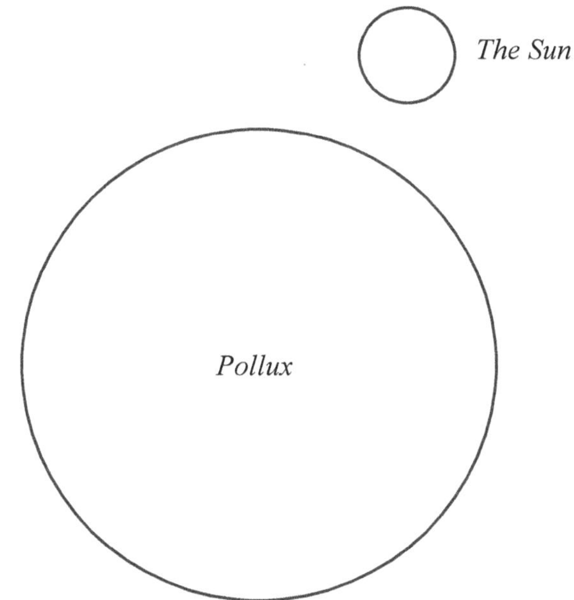

Diagram comparison between Pollux and the Sun

Together with the nearby star of Castor, Pollux is part of the constellation Gemini, the Twins. Sprinkle caster sugar over Pollux is a fun way to remember which star is which.

4. Capella

Capella actually consists of two stars. Capella A and Capella B, which are like each other, both roughly 10 times the Sun's diameter. Over a period of 27 years, as one passes in front of the other, variable brightness can be observed. 'Capella' is the brightest star in the constellation of Auriga.

5. Aldebaran

From our perspective this star appears as the red eye of Taurus the Bull. It is just 65 light years away. Although it appears to be part of the Hyades cluster whereas it is actually nowhere near the main group. The 200 or so stars of that cluster are 150 light years away. In contrast, the nearby Pleiades group of 100 young stars are 380 light years away.

6. Rigel

This is a blue-white supergiant double star, 51,000 times as bright as the Sun. It is **777** light years away, with a temperature of 12,300°C. It is the brightest star in the constellation of Orion.

How many stars in the Winter Hexagon? *I hope by now you are realising that the star you see, is maybe just the tip of an iceberg, so to speak.*

By locating these '6' stars of the Winter Hexagon, you are now armed with guideposts to other stars and constellations. The following diagram shows the location of some of these:

CANIS MAJOR AND CANIS MINOR

This image of the constellation shows the bright star, Sirius, in position as the head of the dog.

Image by Till Credner - Own work: AlltheSky.com, CC BY-SA 3.0,
https://commons.wikimedia.org/w/index.php?curid=20041163

This pair of stars, Sirius A and Sirius B, together make up the brightest star in the northern hemisphere, and the main or alpha star in Canis Majoris. The name Sirius is derived from the Greek meaning 'scorching'. It is one of the nearest stars to us, at a distance of just 8.6 light years away, which is why it appears so bright; it is only twice the mass of the Sun. Sirius B takes 50 years to orbit Sirius A. The constellation represents one of the hunting dogs belonging to Orion. The common name for this star is, unsurprisingly, the Dog Star.

Mirzam is the second brightest star in the constellation and the name means 'herald'. Mirzam precedes Sirius, thus acting as its herald. This blue star is nowhere near that star though, as it is around 492.5 light years away.

Another bright star in the constellation is even further away, a whopping 1800 light years away. This yellow-white supergiant star, Wezen, is at the tail end of the dog. It is 17 times as massive as the Sun yet does not appear as bright as Sirius due to its distance away.

Below Sirius can be found the bright open cluster of stars of M41 which looks as though the stars are arranged in lines.

Canis Minor contains Procyon but very few other stars, and nothing noteworthy. The star, which is comprised of a pair of stars, lies on the celestial equator which makes it useful for locating that, a guidepost if you like. This yellow-white star is 11 light years away and around one and a half times the mass of the Sun.

Once you have located these stars, you will then see Orion striding into view in late autumn as he follows his two hunting dogs.

GEMINI

If you can locate the stars forming the Winter Hexagon, then it is easy to find these two stars forming part of the Gemini constellation.

Image credit Robin Scagell, and a diagram produced by him based on Stellarium Software

Castor and Pollux – the Twins

Caster sugar sprinkled over Pollux helps remember the order of these two stars. According to legend, they are non-identical twins, having the same mother but different fathers. Castor is the son of mortal King Tyndareus of Sparta, while Pollux is the result of the union between their mother and immortal Zeus.

Castor is the less bright of the two. Castor is in fact three stars close together, although they look like two white stars. The 'three' are actually binaries which

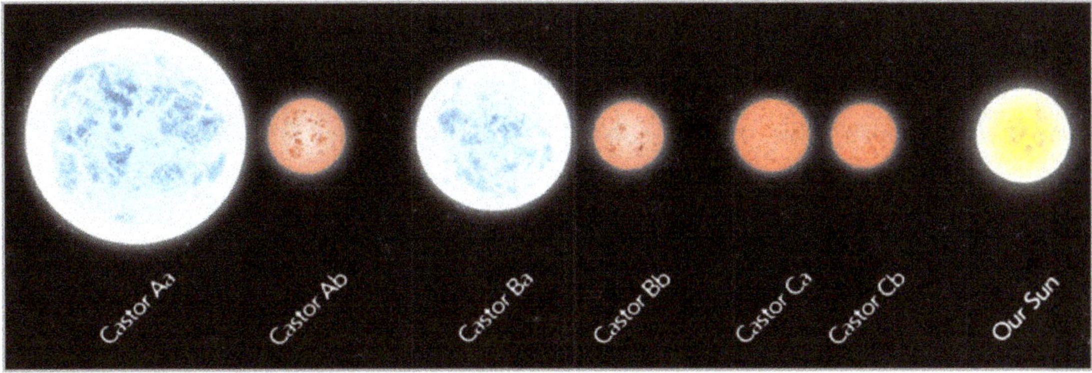

means there are three pairs of two stars, making six altogether. 'It' is 51 light years away.

Pollux and our tiny Sun

Pollux consists of 6 stars, some of which are 1,000 AU apart (*AU = distance from Earth to Sun).* The star is 35 light years away. The Orange or Red Giant is the brighter of the twins, representing the immortal offspring from the union between their mother and Zeus when disguised as a Swan. This is a star that has finished fusing hydrogen in its core and is now fusing other lighter elements into heavier ones. The star has a temperature of 4,627°C. Recently, astronomers discovered that the star has an extrasolar planet, or exoplanet.

M35 Open Cluster is down by the 'feet' of Castor with around 100 stars in curving chains. It is visible through binoculars. In the same area as M35 are two more open clusters, NGC2158, and IC2157.

The **Geminids Meteor Shower**, between 13-14 December each year, is one of the most prominent in the year. There can be as many as 100 meteors per hour. An occasion for wrapping up warmly to stay outside for an hour or so, before coming inside for a welcome hot chocolate. It is these treats that make astronomy special!

AURIGA

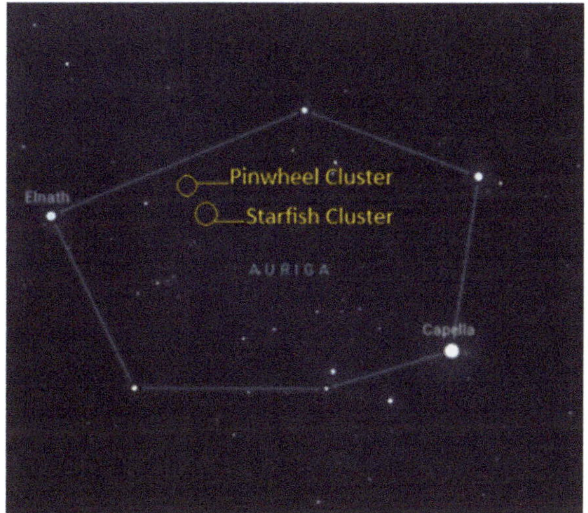

Credit: Stellarium Web (extract from original image)

This contains Capella, which, as one of the early stars to be seen as dusk falls, is one of the main navigation stars.

Capella is part of the Winter Hexagon asterism. Being right overhead during winter, it is an easy star to spot. Capella (Alpha Aurigae) is the main star in the Auriga constellation and is the sixth brightest star in sky. It is actually a pair of yellow giant stars which are 42 light years away.

M36 Open Cluster 'Pinwheel Cluster' contains around 60 stars. Visible in binoculars, this is a cluster similar to the Pleiades. The cluster is 14 light years across and 40,000 light years away (*one light year is 6 million million miles, roughly!*).

M37 is a rich Open Cluster of about 150 stars, about 4,500 light years away. The cluster is about 20–25 light years across. This cluster was missed by French astronomer Guillaume Le Gentil (*oh dear! he was not a lucky chap!*) when he **re**discovered M36 and M38 in 1749. All three had already been identified a century earlier by an Italian astronomer.

M38 'Starfish Cluster' This is a scattered Open Cluster of about 100 stars, which is visible in binoculars. It is 3,400 light years away and 25 light years across, similar to M37. A smaller cluster, NGC1907, lies to the south.

There are also a couple of variable stars within this constellation. One brightens and lessens in brightness every 3.7 days while the other is comprised of a star which is eclipsed by another with the eclipse lasting 27 years.

Auriga

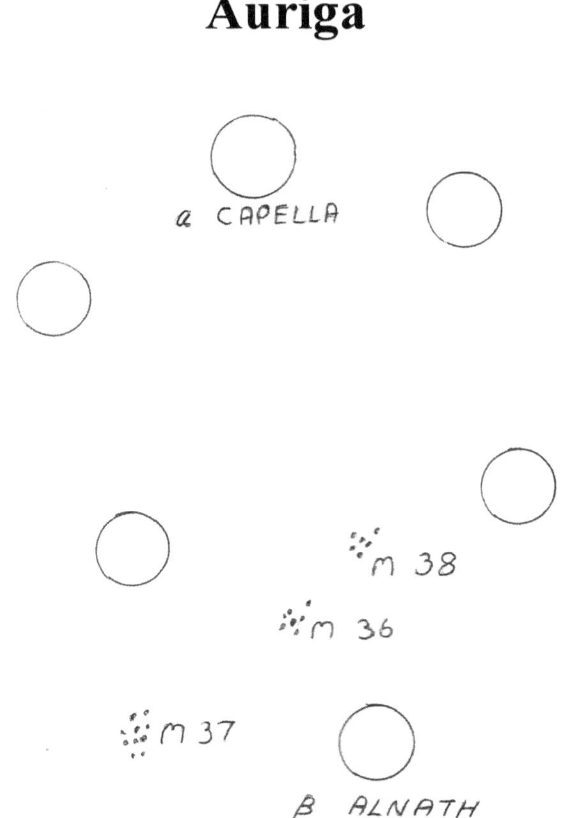

The star Alnath is also shared with the constellation of Taurus. Although it is marked on Auriga on some star maps, it is now officially part of Taurus. This is useful to know because by locating Alnath, you can then locate two constellations from that star, or by locating one constellation, you can then locate the other.

TAURUS

Credit: ESA Akira Fujii

This constellation offers the delights of seeing very old stars and a fairly new cluster of stars without too much difficulty. The Pleiades is a well-known cluster as it is easily visible to the naked eye.

Aldebaran

This bright red star is one of the early ones to appear and is therefore another good one to learn as you will then be able to use that knowledge to find other stars. It is the red eye of Taurus the bull. It is an orange giant (a star nearing the end of its life), appearing in front of the Hyades. Although it looks to be part of that cluster, it is quite separate. The Hyades are 150 light years away, with Aldebaran much closer to us at half that distance. An old red giant of a star, at 6.605 billion years old, while our Sun is 4.603 billion years old.

Pleiades (M45)

The Pleiades are commonly known as the Seven Sisters, although somewhat inevitably there are a quite a few more than 7… around 750 stars comprise this group. It is one of the brightest and youngest star clusters in the sky, having formed around 100 million years ago, about 444 light years from Earth. From side to side the group

spans 13 light years, or about halfway from Earth to the bright star of Vega. Behaving rather like a school of fish, the stars move together as a gravitationally bound swarm through space.

Hyades

This is an open cluster of about 200 stars. At 150 light years away, it is the nearest cluster to us. It is believed to have formed 800 million years ago, making it quite an old cluster. You will see this cluster of stars around the red star of Aldebaran.

Credit: NASA, ESA, and STScI.

M1 The Crab Nebula is a remnant of a supernova (1054 AD), 6500 light years away. It is just visible as an elliptical hazy patch through binoculars.

The star, Alnath, at the tip of one the 'horns' of the bull, has a history of being shared with the constellation Auriga. The name is derived from the Arabic for 'butting'. It is the second brightest star in Taurus and is 134 light years away.

ORION

When Orion strides into view, you know it is a sign that winter is well under way and with it, clear, dark skies (weather permitting!) due to the decrease in atmospheric disturbance. When the air is colder, the atmosphere holds much less moisture compared to the summer months. The additional moisture probably hinders the passage of starlight. Personally, I feel sad somehow when I see Orion high in the sky to the west as we head into spring, a visible reminder of the passing of time perhaps?

Light Years

Before taking a look at the stars in Orion, it is helpful to have a broad understanding of what a light year is in order to appreciate the vast distances involved when looking at this constellation in particular, reasons for which will become apparent. A light year is the distance light travels in a year. The speed at which it travels is 299,792 km/186,000 miles per second; it takes just over 8 minutes for light to reach us from the Sun. (Distance from Sun to Earth is around 150 million km/94 million miles) So, a light year is quite some considerable distance… 10 trillion km/6 trillion miles… or

10,000,000,000,000 km/6,000,000,000,000 miles

(trillion = million million)

What is a star?

A star is formed out of a collapsing cloud of interstellar matter. When the pressure and temperature become so high that nuclear reactions start, that matter becomes a star. Hydrogen is converted into helium. The force of its gravity acts as a counterbalance to the nuclear reaction and keeps the star in equilibrium. Large stars are pulled towards the base of a nebula by gravity.

Nebulae are star forming regions. Stars are formed within giant gas clouds (nebulae) where dust swirls deep within them. Pockets of gas are formed which grow so big that gravity pulls more gas into the pocket. Then gradually, the pocket of gas begins to contract, and it slowly warms up. When the pressure and temperature become so high, nuclear reactions start and the matter becomes a star.

Stars range from a mass of 0.08 to 100 (1 mass = the Sun), too small or too big and they are unsustainable.

****

Now onto the excitement of Orion…

M42 Orion Nebula

It is worth visiting an observatory to view this feature through a telescope. If you have 10x50 binoculars you will see quite a few of the features or, at least, you may catch a glimpse of the fuzzy blob with the naked eye. It is an excellent viewing target for January and February.

The Orion Nebula is a major star forming region, with four central stars in the foreground forming the Trapezium feature. Recent research has shown stars being formed here. It is **1344** light years away. An interesting project is to make a 3D model of Orion to illustrate just how far the Nebula is from the main stars of the constellation. The whole area of M42 extends several hundred light years across, covering a much larger area than what is visible from Earth.

This image was released on 29 February 2012. Combining far-infrared observations from the Herschel Space Observatory (145,000 km/900,000 miles from Earth) and mid-infrared observations from NASA's Spitzer Space Telescope, this image shows newly forming stars surrounded by remnant gas and dust in the form of discs and larger pockets.

Binoculars or a telescope are appropriate for appreciating the Trapezium within the Nebula (20x60 binoculars on a tripod would offer the best view). It is a multiple star system at the heart of the nebula, containing hot young stars. Within the Orion Nebula is a glow of about 30 light years across from ultraviolet radiation from the new stars forming within it.

This region also contains variable stars, i.e. ones that fluctuate in brightness over time. Visible through binoculars is a dark lane running east from the Trapezium. This feature, known as the Fish's Mouth separates parts of the Nebula with the Trapezium emerging in the centre. The **Horsehead Nebula** is a popular image for posters although not visible except by long exposure telescopes. It lies deep within the Orion nebula region, 1,500 light years from Earth.

Our view from Earth…

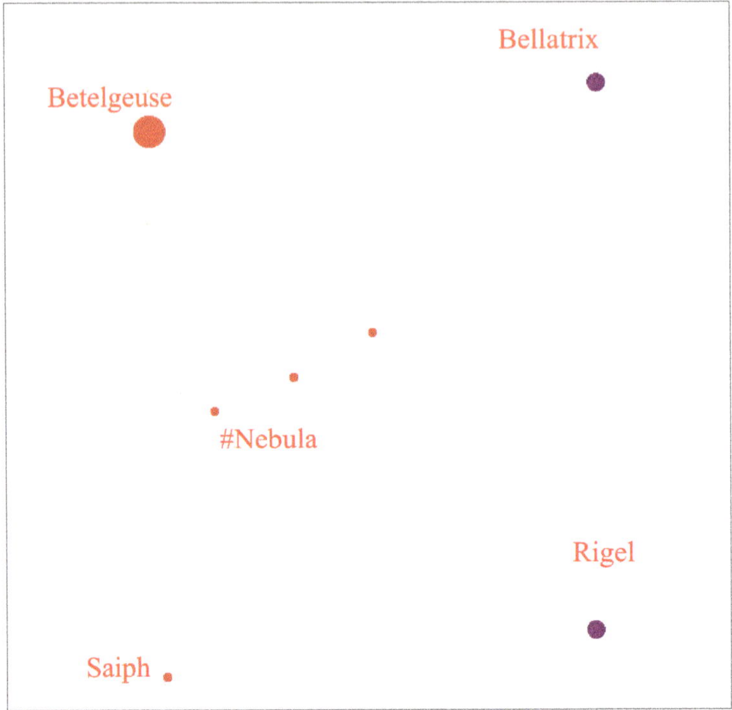

…if the stars could be seen from a sideways perspective in space,
the constellations would look very different.

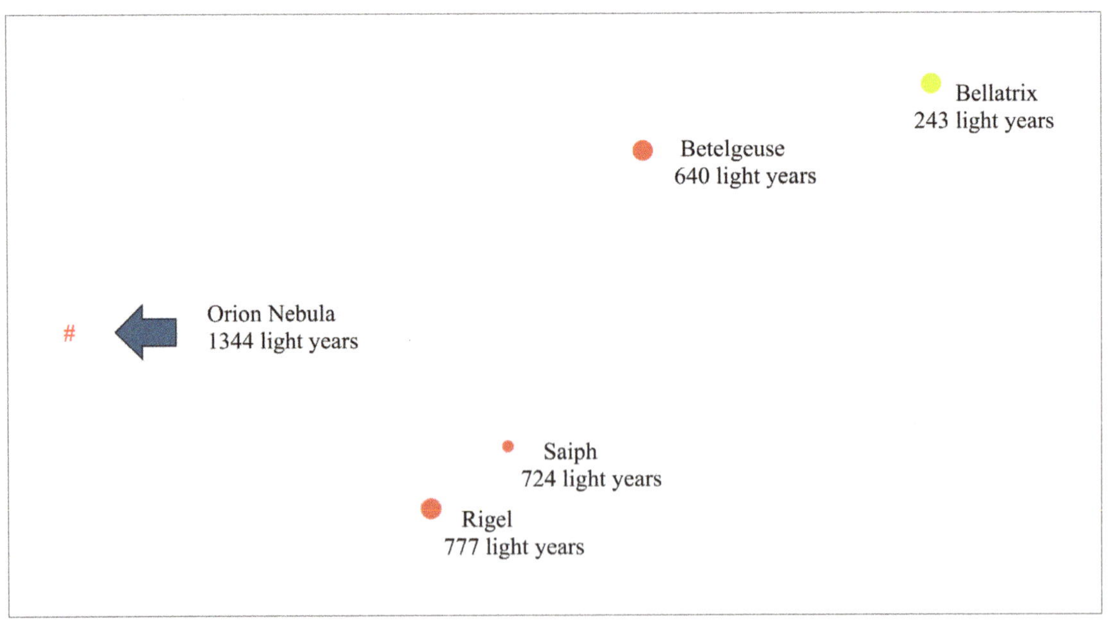

Betelgeuse

This is a red Supergiant, a star towards the end of its life, it is 1000 times larger than the Sun; if it was where our Sun is, it would extend out as far as Jupiter! **640 light years away**, and 59,000 times as bright as the Sun. It is only a few million years old though. It is a variable star, and during 2019 were signs of it dimming and brightening again.

Bellatrix

This is a younger star, a blue giant which is actually a pair of stars. It is **245 light years away**, and 6,000 times as bright as the Sun.

Rigel

This young double star is a blue-white supergiant – double star. It is **770 light years away**, and 51,000 times as bright as the Sun. Temperature is 12,300°C.

Saiph

This is **650 light years away**, and 19,000 times as bright as the Sun. Again, this is an enormous star, with 14–17 times the mass and over 11 times the diameter of the Sun. It is another supergiant star that has exhausted the supply of hydrogen at its core. This one is losing mass at a rate equivalent to the Sun's mass every 1.1 million years. Large stars such as Saiph are destined to collapse in on themselves and explode as supernovae or exploding stars.

Stars in Orion's Belt

Alnitak – a pair of stars with a white companion. This star is the dimmest of the belt stars and is 900 light years away. It is only 7,000 times brighter than the Sun even though it has the same mass and surface temperature of the mighty Mintaka.

Alnilam – blue-white supergiant at the centre of the Belt. It is 1,360 light years from Earth. It is the brightest of the Belt stars at 18,000 times brighter than our Sun. The mass is about the same as its fellow stars at 20 solar masses. The surface temperature is the coolest of the three at only 50,000°C. It is five times further from us than Bellatrix.

Mintaka – a blue-white supergiant with a companion. It is 915 light years from Earth and is the westernmost star in the belt from our perspective on Earth. The star is 10,000 brighter than our Sun and has a searing hot surface temperature of 60,000°C.

*****

Norman Lockyer and Spectroscopy:

The Orion nebula was studied by Norman Lockyer using the lens that is in the 'Lockyer' telescope at the Norman Lockyer Observatory. In the Sun, 600 million tonnes of hydrogen is converted into 596 tonnes of helium every second. Second most abundant element in universe – after Big Bang 25% helium, 75% hydrogen.

French astronomer Pierre Janssen and British Norman Lockyer, almost simultaneously in the late 19th century, developed a technique for examining the spectrum of a solar flare, for which the French Academy of Sciences awarded each a commemorative medal honouring them both. When light is 'split' coming out of an object, the colours equate to certain chemicals. This process is known as spectroscopy. Images of these colours are shown in lines (Fraunhofer Lines).

However, one of the lines did not correspond to known elements and Lockyer asserted that it was due to some hitherto unknown element, which Lockyer named helium. He later found the same element to be present in the Orion nebula. It was to be another 20 or so years before helium was identified on Earth.

LEO

The outline of stars for this constellation resembles a crouching lion, with the head appearing as a giant backward question mark in the sky. It is worth trying to locate this one as it gives you a starting point for enjoying the delights of spotting galaxies out beyond the arms of our own galaxy. To Leo's left is the huge constellation of Virgo which contains many more galaxies than it does stars, but more of that one later.

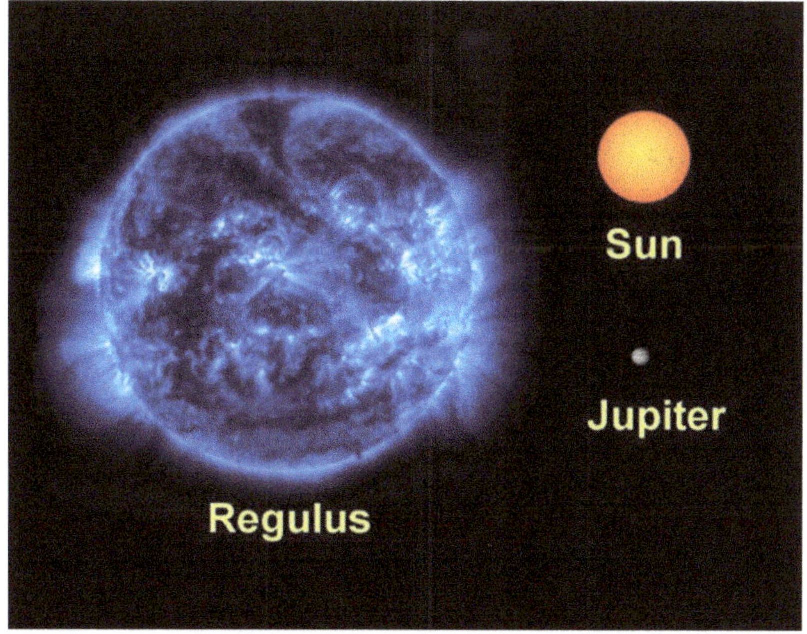

Regulus and the Sun size comparison

In the far reaches of deep space, many galaxies can be viewed as fuzzy blobs through binoculars or spirals through a telescope. Either way, it is possible to see galaxies other than our own and to contemplate what the Milky Way Galaxy looks like from afar. During springtime and autumn, due to the tilting of the Earth, we are able to look out across the arms of the Milky Way into the depths of the area surrounding Leo.

Some specific features in Leo to seek out are described here:

Regulus
This, with a companion star, is visible through binoculars. Forming a large triangle with Spica and Procyon, it is found in the opposite direction to Arcturus; instead of 'arc to Arcturus', follow the pointers in the Plough in the opposite direction. This star lies on the ecliptic, 79 light years from Earth.

Regulus is a multiple star system consisting of at least four stars. The main star is a blue-white star 300 times more luminous than the Sun. It is a fast-rotating star with its companion stars taking 130,000 years to orbit it. These stars are around 4,000 astronomical units (*the distance of the Earth from the Sun*) away from Regulus.

Leonids Meteor Shower
Around 17 November, this meteor shower is worth trying to see. It comes from the comet Tempel–Tuttle. This is a comet with an orbital period of 33 years which passes through the constellation of Leo at this time of year. It was independently discovered by Wilhelm Tempel on 19 December 1865, and by Horace Parnell Tuttle on 6 January 1866.

SOME GALAXIES

M65 A spiral galaxy

M66 A spiral, although looks cigar-shaped from side on

M95 and M96
These two can be seen as faint smudges on a dark night

These are around 35 million light years away... so it's not surprising they are a bit faint! And also, whilst looking at these distance smudges, reflect that, from them, the Milky Way Galaxy will also look like a fuzzy smudge.

The diagram below shows the location of the main stars and galaxies around Leo:

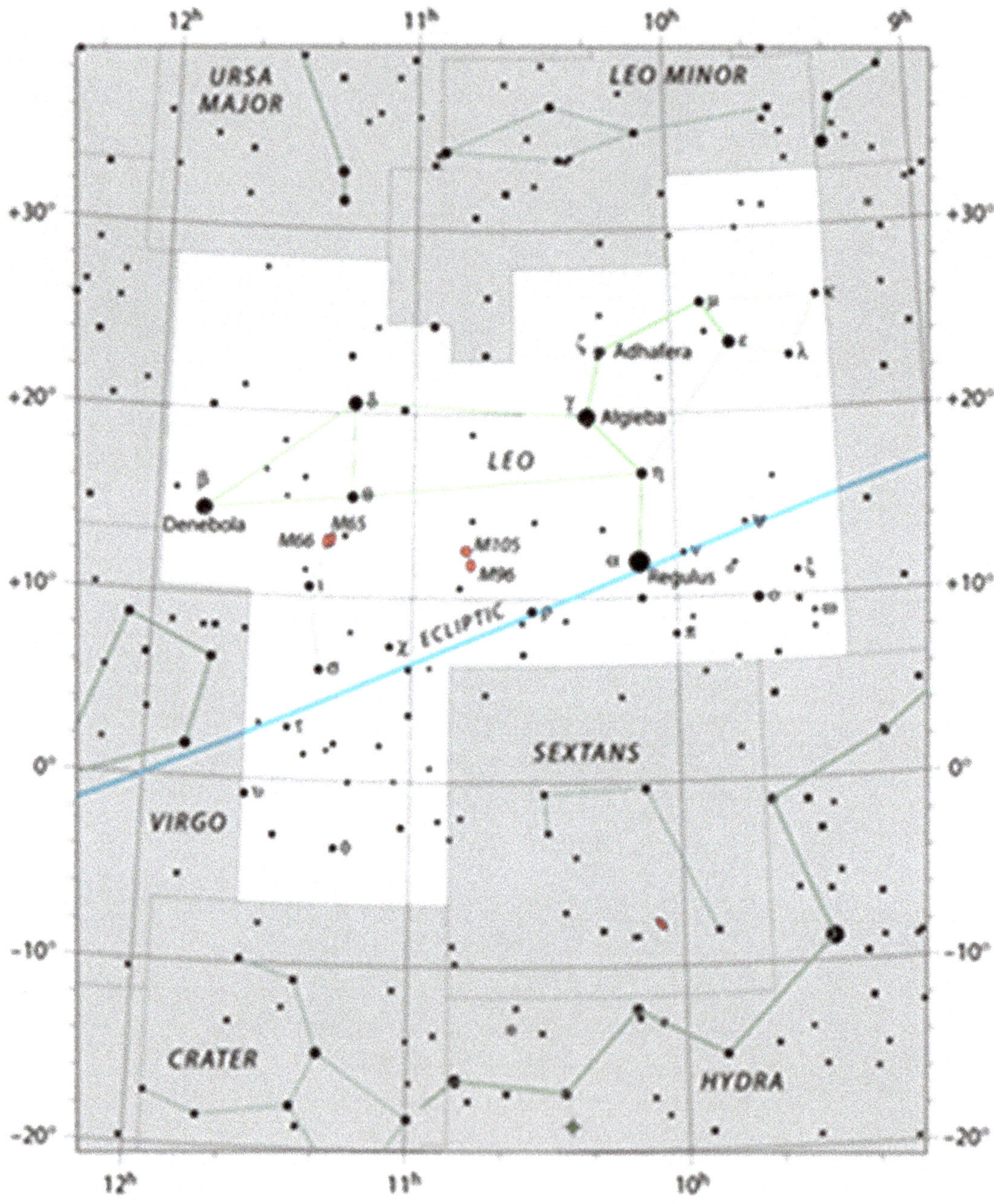

VIRGO – HARBINGER OF SPRING

Credit: NASA, ESA, and the Digitized Sky Survey 2.
Acknowledgment: Davide De Martin (ESA/Hubble)

If Leo has whetted your appetite for galaxies, then Virgo is not going to disappoint. Unbelievably, it contains many more galaxies than it does stars. Virgo encompasses our own Local Group of galaxies, 2,000 galaxies each one many thousands of light years across, between 10,000 and 100,000 light years diameter. The centre of the Virgo Cluster is 60 million light years away. The cluster is concentrated in a patch of sky in western Virgo, some 8-10 degrees wide. The Virgo Supercluster includes our own galaxy plus the galaxies in Leo in addition to those in Virgo.

With the tilting of the Earth, the Milky Way lies along the horizon during springtime, making it easier to see deep sky objects beyond, such as the galaxies in Virgo during April – June although one runs into the difficulty of the ever-lightening evenings!

Spica

This star is one of the navigation stars used by ancient mariners. From the Plough **arc** down to **Arcturus** and then you take **a spike down to Spica**. It is a blue-white eclipsing binary star. It is 250 light years away from Earth. This constellation is a harbinger of spring and was used by previous generations as a signal for planting time. The name Spica means 'ear of wheat'. It is the second largest constellation (Hydra is largest) and is seen in both northern and southern hemispheres and overhead at the Equator. With an image of a maiden carrying a sheaf of wheat, Babylonian, Greek and Roman myths and tales all relate Virgo with agriculture. This is not surprising as the northern hemisphere spring is synonymous with planting time. Interestingly, in the southern hemisphere, Aborigines saw an emu in the sky and used its orientation to decide whether it was time to hunt for emus or collect their eggs.

Porrima

It is an interesting binary star in that its variability means the stars come close together before moving away again. They are the furthest apart during 2020 when it was possible to view both stars through binoculars! It takes them 172 years to complete their cycle. The stars are just a mere 35 light years away.

70 Virginis

Although difficult to observe (which is counter to the point of this book!), this star is worthy of being included because of its historical interest. This star has one of the first known exoplanets, with one confirmed planet 7.5 times the mass of Jupiter. It is also an interesting object in that it pulsates over the course of a year from 6th to 10th magnitude – quite a huge range in size and brightness!

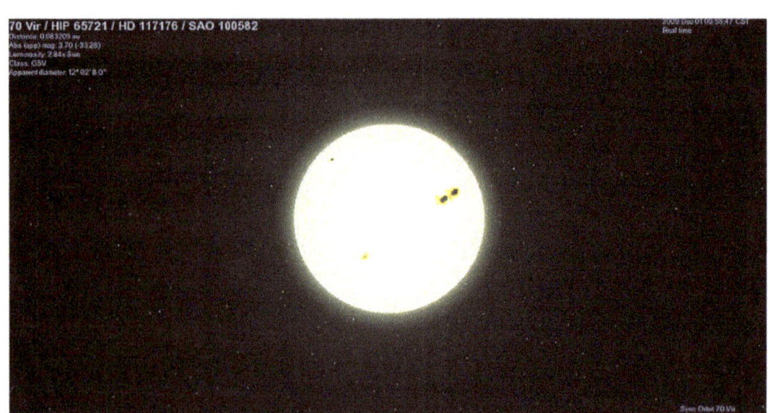

70 Virginis: Chris Laurel - Screenshot, GPL,
https://commons.wikimedia.org/w/index.php?curid=864427

****

A selection of galaxy highlights in Virgo

M84 elliptical galaxy. There are two jets of matter which have been observed coming from it. It forms a pair with M86 in the same field.

M86 displays the highest blue shift (Doppler effect of objects coming nearer) of all Messier objects, as it is approaching the Milky Way at 244 km (approx. 152 miles). This is due to its falling towards the centre of the Virgo cluster from the opposite side, which causes it to move in the direction of the Milky Way.

M87 was the first galaxy to be discovered with a black hole in it. It is a strong source of radiation, the brightest radio source in Virgo and also known as Radio Source Virgo A.

M104 Sombrero Galaxy is an unbarred spiral (about one third of galaxies are like this) located 28 million light years from Earth. This galaxy is not part of the actual Virgo cluster but is 20 million miles closer to us. It is in the 'Y' of Virgo and is an active galaxy. A jet of energetic plasma (*gas containing charged particles)* originates at the core and extends outward at least by 5,000 light-years. It has a bright nucleus, an unusually large central bulge, and a prominent dust lane in its inclined disk. The dark dust lane and the bulge give this galaxy the appearance of a sombrero.

Hubble image of Sombrero Galaxy

Hubble space telescope image of galaxies

MILKY WAY

This photograph was taken by NLO member, Alan Green who said, "I took some photos about four weeks ago and just got around to looking at them. I was just using a canon 800D with an 18x55 stabilizer lens with an Omegon clock drive that tracks the night sky on a camera tripod. I took 10 images at 30 seconds then stacked them." (Sept 2020)

On a clear dark night, you can see a swathe of stars sweeping across the sky in a broad band. This is just one arm of the spiral galaxy in which our solar system resides. Ancient Greek astronomers thought the band resembled spilt milk, hence its name via Lactea or milky circle, from which is derived the name by which our entire galaxy is known. It has also featured in many other cultures as a road or lane.

And the following illustration shows the location of Earth:

Interestingly, the Milky Way was initially thought to contain all the stars in the Universe, so how was the band of the Milky Way known to be an arm of just one galaxy?

As early as 1750 it had been suggested that the Milky Way was a large rotating disc of stars, following on from Galileo's observations that the solar system contained countless stars. William Herschel, in 1784, concluded the Earth to be at the centre of a galaxy, having estimated the shape and size by surveying individual stars.

However, the existence of other galaxies was only proven in 1924 by Edwin Hubble (whose name lives on with the Hubble Telescope). At the time of Hubble's research, there was only one galaxy known, that of the Milky Way, of which all the stars were thought to be part. No one knew how many stars there were. At that time, female 'computers' analysed photographic plates and one, Henrietta Swan Leavitt, noticed that some stars pulsated regularly. She devised a method for comparing the relative magnitudes of such stars in order to work out the relative distances between them. Hubble used her results to analyse and observe a blurry mass within the Andromeda constellation. This resulted in his 1923 groundbreaking paper in which he positively identified M31 as a mass of stars, or galaxy, not a gas cloud, and in 1924 it was estimated that it was at least 900,000 km away (it is actually 2.5 million km away). Following this he went on to develop a classification system based on the shape and structure of different types of galaxies.

Similar research was carried out in Lund University by Swedish astronomer, Dr Knut Emil Lundmark PhD. In 1919 his assistant, hydrographic engineer, O'Jahnke produced a map of the Milky Way under Lundmark's supervision.

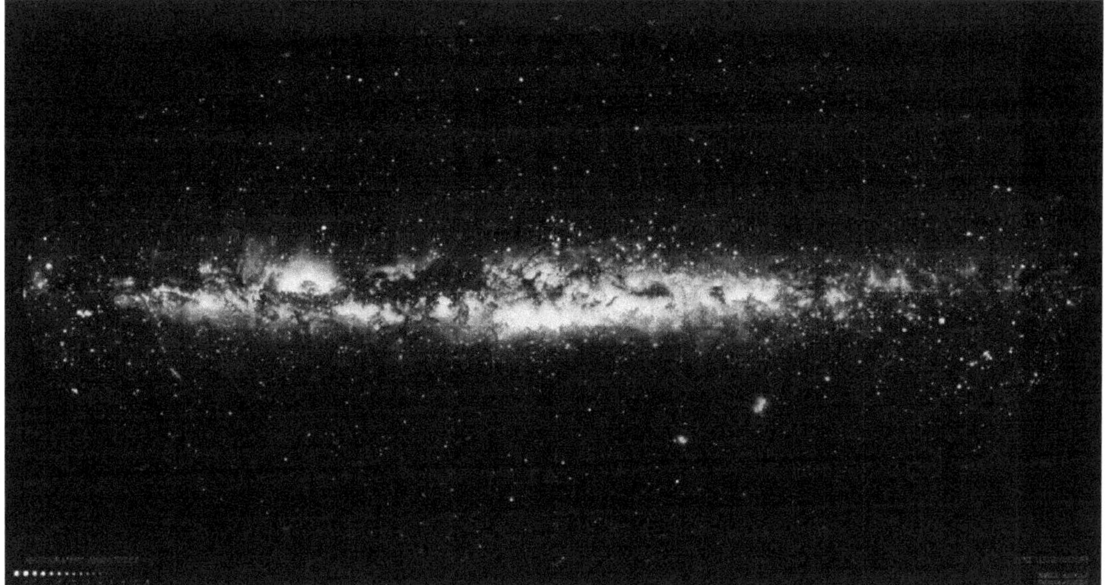

Image courtesy of Lund University

In the 1950s the Milky Way was confirmed as a spiral galaxy with the identification of the Perseus and Orion arms, and in 1997 that the centre of the galaxy contained a massive black hole. Scientists continue to map the stars and galaxies with the most comprehensive to date being the Two Micron All-Sky Survey, a collaboration between the University of Massachusetts at Amherst and the NASA's Infrared Processing and Analysis Centre. So, the name Milky Way refers to both an entire galaxy as well as that lane of stars you see on a clear night.

*****

Returning now to the naked eye observation of that milky lane of stars, if you look closely, you will see that the bands of stars are separated by lanes of dust. This interstellar dust comprises material shed by older stars and material which will make new stars, hydrogen, and as-yet-to-be identified matter, known as 'dark' matter simply because it is unknown not because it is dark. When you look up to those stars, you are looking out from within our galaxy across the plane of the galaxy into space. And when you look towards the constellation of Sagittarius, you are looking towards the centre of the galaxy.

AND SO WE'RE BACK TO THE SUMMER TRIANGLE…

Now that you have gained a broad understanding of where some of the constellations are in relation to the ecliptic and the main navigation stars, you are now well-prepared to further your knowledge using the many apps and books that are available. See reading list for some suggestions. There are a total of 88 constellations to be explored, as well as many other interesting features in the night sky, such as comets, meteor showers and all those star clusters, ancient and not so ancient. And of course, the Moon and the Sun.

Author's note: I would say 'most' rather than all!

Clear Skies and Happy Viewing!

*****

Part 3

Planets

MERCURY

By NASA/Johns Hopkins University Applied Physics Laboratory/Arizona State University/Carnegie Institution of Washington - https://photojournal.jpl.nasa.gov/catalog/PIA11364, Public Domain, https://commons.wikimedia.org/w/index.php?curid=83618472

This planet is **very hot**, **very fast** and **very bright**: ancient cultures were aware of the speed at which it goes around the Sun, each calling it the Messenger of the Gods. Mercury is its Roman name. On the surface are long cliffs stretching out far, far into the distance with equally long shadows cast by the Sun when low on the horizon.

This tiny planet is a challenge to observe due to its proximity to the Sun. It is visible when it is either side of the Sun or in transit. It is possible to catch a glimpse of it just before daybreak before it sinks below the horizon (be careful not to blind yourself by the rising Sun though) or just before sunset. Another opportunity to observe its presence is when it transits the Sun. We are in the fortunate position of being able to look at Mercury and Venus when they pass across the face of the Sun, or transit, because of our position of being third away from the Sun. It cannot be emphasised enough though, **not** to look through any equipment without proper filters. It is best left to experts. These transits are well-publicised and covered in newspapers and astronomy magazines and online.

When you are based near the coast as I am, observing tends to be tinged with frustration. It proved impossible to view the transit of Venus in 2012 due to sea mist and cloud. However, it was almost as exciting to see the event on a friend's laptop... but I digress. Mercury, being tiny, is even more of a challenge for an amateur in normal circumstances let alone when in transit. Early one morning, we (Mum, Dad, and enthusiastic young daughter) set off to a nearby hill over-looking the sea. We just about saw a tiny planet in the moments before sunrise. The view from the hill gave us a good angle from which to safely spot the planet before the dangerous time of sunrise.

*****

The search for exoplanets uses the same techniques for transits on distant stars i.e. there is a dip in the light being emitted from a star when an object (planet) passes across it.

The Mariner and Messenger missions to Mercury have given an insight into its composition and appearance. It has a large iron core and is a rocky planet. Images from the Messenger missions to Mercury between 2011 and 2015 showed ice at the poles. This ice may be result of a comet impact and has stayed due to those parts of the craters remaining in the shadows. The surface is heavily

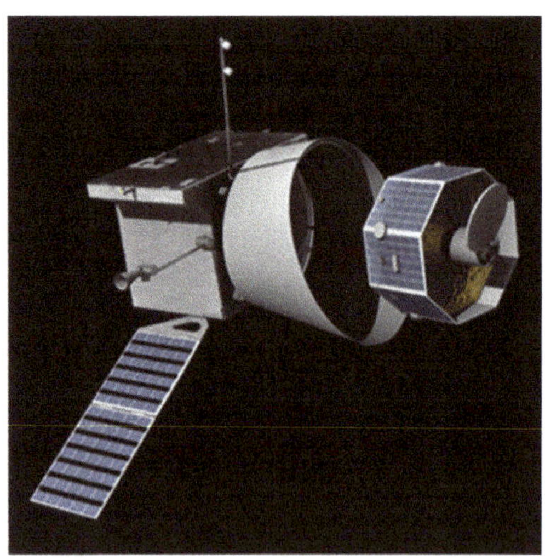

pockmarked with many impact craters. Mercury lacks a protective atmosphere so is vulnerable to objects hurling themselves at the surface. The largest crater is the Caloris Basin which is about 1,525 km/ 950 miles across and is ringed by mile-high mountains. Further investigation into the planet is underway with the latest technology. A probe was launched in 2018 and will reach the planet in 2025; Bepi-Columbo, a joint project between Europe and Japan, will use special techniques to slow down in order to avoid the gravitational pull of the Sun to successfully orbit Mercury. At present, it is the least explored planet in the solar system.

Despite being so near to the Sun, it is not the hottest planet. It has a wide temperature range from 430°C by day to -180°C at night. It has a fast orbit around the Sun of just 88 Earth days due to its proximity to the Sun, but its very slow rotation period gives the planet very long days and very long nights. It takes a massive 58 days to make one rotation. For comparison, our moon takes 28 days to rotate and orbit the Earth, while we spin on a 24-hour cycle while taking 365/6 days to orbit the Sun. The planet is not tilted like the Earth and is relatively 'upright' which means it has no seasons. The combination of its speedy elliptical orbit and slow rotation means that some parts of the planet experience a double sunrise and a double sunset on occasion. The Sun would appear to someone on Mercury to be around 3 times the size as viewed from Earth.

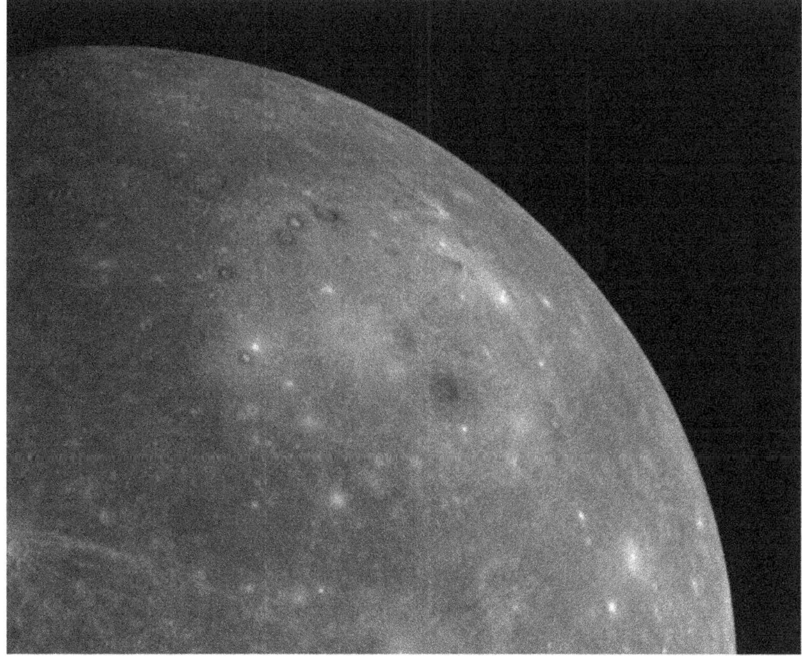

Image by NASA/Johns Hopkins University Applied Physics Laboratory/Carnegie Institution of Washington - Image: MESSENGER first photo of unseen side of mercury.jpg, Public Domain, https://commons.wikimedia.org/w/index.php?curid=3405283

VENUS

Artist's impression by John Meacham

Known as Earth's Evil Twin, Venus is a tricky object to observe. Yes, it can easily be seen in the sky as the evening or morning, but it is shrouded in a thick atmosphere which needs a special filter to actually observe anything of interest. Its thick clouds reflect the Sun's light very well which is why it is so bright. The limited visibility is due to its proximity to the Sun, as with Mercury, although it is easier to spot and often called the Evening or Morning Star.

High in the sky it appears as a pretty, pearly planet hence its name of 'Venus' after the goddess of beauty. The planet rotates backwards as it orbits the Sun. Venus takes longer to complete a single rotation than it takes to orbit the Sun (224 days to orbit, 243 to rotate). This means the Sun rises once every 118 Earth days, a double sunrise in effect as the length of the day is so long, and the Sun rises in the west and sets in the east. That means the Sun rises two times during each year on Venus, even though it is still the same day on Venus! And because Venus rotates backwards, the Sun rises in the west and sets in the east. The clouds and the atmosphere on Venus rotate about 60 times faster than the planet itself. From those clouds fall lethal sulphuric acid rain.

What does Venus and coca cola have in common? They both contain massive amounts of carbon dioxide! Carbon dioxide makes up 96% of the atmosphere on Venus. The rest is made up of nitrogen and other gases. There has been no indication of oxygen… although there are traces of water vapour, argon and sulphur dioxide. The presence of granite on Venus indicates that there was once water on the planet as granite needs water in its formation.

The planet demonstrates an extreme example of what can happen with a runaway greenhouse effect. The pressure is 90% greater than on Earth. Both are in that marvellous habitable zone! The temperature is over 400°C on the surface which is the result of a runaway greenhouse effect. The thick atmosphere is 80km/ 50 miles deep with three cloud layers comprising mainly sulphuric acid. The planet was probably similar to Earth initially but may have had atmosphere blown away by something.

The surface of Venus, as seen by various flybys from the early Mariner programme in the 1960s to the later Mariner programme, is a rocky, orangey landscape. Magellan, during the 1980s programme, mapped the surface of the planet which showed mountains some 11 km/7 miles high and lava fields from numerous volcanoes. Between 1973-83 the Russian Venera series actually landed on the surface. Venera 13, a Soviet spacecraft, was the first lander to transmit colour images from the surface of Venus. Although the spacecraft was designed to last about half an hour in the harsh Venusian environment, it transmitted data for more than 2 hours after its landing on 1 March 1982. No lander has ventured to the surface of Venus since the 1980s. Such landing craft were specially designed to withstand the intense pressure. If a human was to walk on the surface, they would be squashed flat instantly from the intense pressure. Although the planet sits with in the Goldilocks Zone for habitable planets, it has such a thick, poisonous atmosphere, life is impossible. In 2020, there was the possibility that phosphine had been found within the deep clouds which are an indication of a biological life form although further research could not confirm this.

A depiction of the volcanic surface of Venus by Alison Young

Appearing as a slightly larger dot than Mercury when it passes in front of the Sun, Venus can be seen when it is in transit across the surface of the Sun. In 2007 and 2012 this was apparent from the northern hemisphere. As mentioned earlier, I went along to the Norman Lockyer Observatory early in the morning in 2012 only to be thwarted by sea mist and cloud. I viewed it via a friend's laptop. Frustrating to say the least! More success was had in 2007 when my daughter set up her telescope on the front step of the house and used paper to reflect the poor image.

However, we were much more fortunate than the French astronomer, Guillaume Le Gentil who travelled to India to witness it. He was ill so missed the first transit (they come in pairs), and it was cloudy for the second. In the intervening years he voyaged down to Africa and back. When he eventually arrived back home to France it was to find he had been declared dead and his wife remarried. And he hadn't even seen the transit.

Drawing of Venus - John Meacham

This is a drawing made by John Meacham as viewed through a 5" reflector telescope using an orange Wratten filter in 2012.

EARTH

Credit: Alison Young

Earth is the third planet away from the Sun in the Milky Way Galaxy. The human race sometimes needs reminding of the fact that Earth is indeed a planet with finite resources. However, it is a planet like no other in our own solar system, supporting life within its oxygen rich atmosphere. It contains a multitude of life forms and is firmly within the fabled Goldilocks Zone, along with Venus and Mars, incidentally. Earth has not had its atmosphere 'blown away' by the solar wind (Mars) nor has an atmosphere so thick that it heats up beyond imagination (Venus). It also is blessed with a plentiful supply of oxygen and water, essential for life as we know it.

There are four main elements to this planet: air, water, land, ice. Humans have managed to conquer most areas of the planet except for the deepest oceans. It is harder to reach the bottom of the Marianas Trench than it was to reach the Moon. This trench is so deep the water would completely cover Mount Everest. There are a multitude of books about Earth, and I would urge you to pore over the pictures of those which celebrate its natural wonders. I am fascinated by both physical and human geography so any travel book would be of value in helping to understand the relationship between land and people.

Earth is referred to as the 'Blue Planet' because it appears as a small blue ball when viewed from space. Other terms include 'Spaceship Earth' and 'Fragile Earth' in deference to the fact that we are dependent on the delicate balance of the Earth's atmosphere to protect us from the Sun's harmful ultraviolet solar radiation as well as modifying the temperature changes during day and night-time, **and** to recognise that we are all inhabiting a common home, contrary to the machinations of politicians, as we travel through space.

Thanks to the Earth's 23-degree tilt relative to the Sun, most of the Earth's surface experiences seasons, due to the increase or decrease in sunlight reaching that part of the planet.

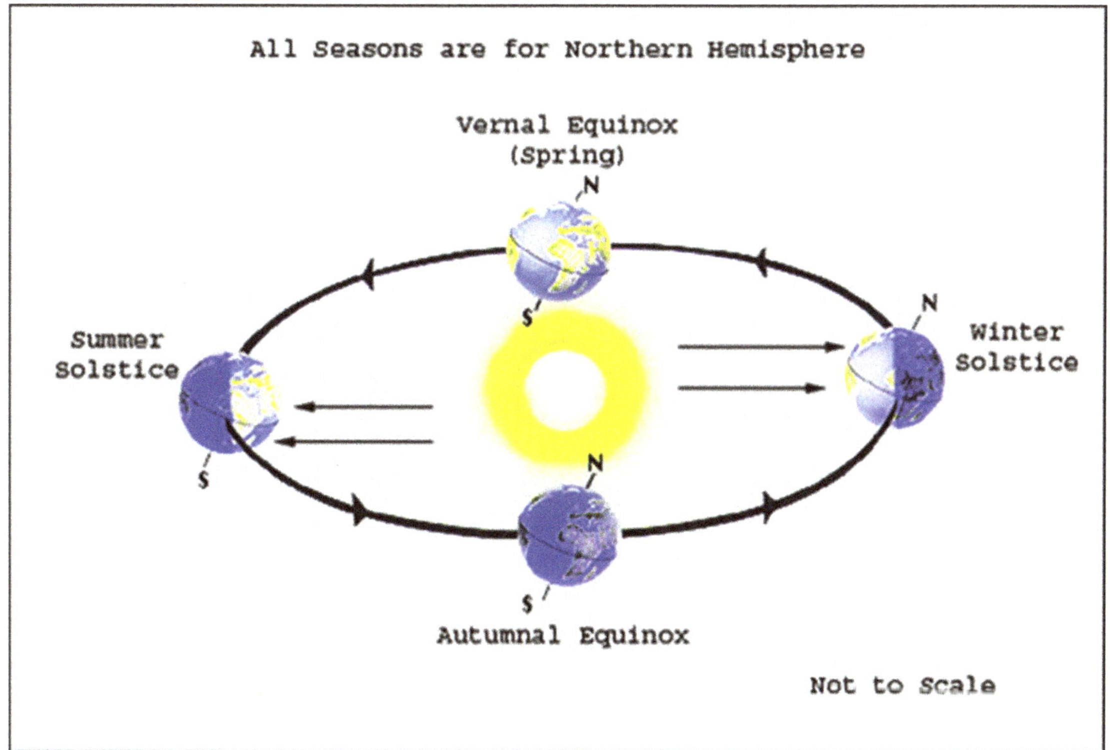

The Moon and Tides

The greatest effect the Moon has on the Earth is in relation to the water on the Earth. High and low tides on Earth are caused by the Moon pulling the water which causes it to bulge thus creating tides. If the Moon was not there, there would less tidal activity. The gravitational pull of the Sun is weak due to the distance the Sun is away from Earth. With the addition of the Moon's gravitational pull, the tides range from very low to very high according to where the Moon is in relation to the Sun. When both are pulling from the same direction, there are high spring tides throughout the year which cause much damage from erosion and movement of sand. Sea stacks are formed, collapse over time, cliffs crumble into the sea and great spits of sand form on coastlines.

In the diagram below, the shaded 'bulges' at each end of the Earth show how the water is pulled in the direction of the Moon and the Sun at high spring tides:

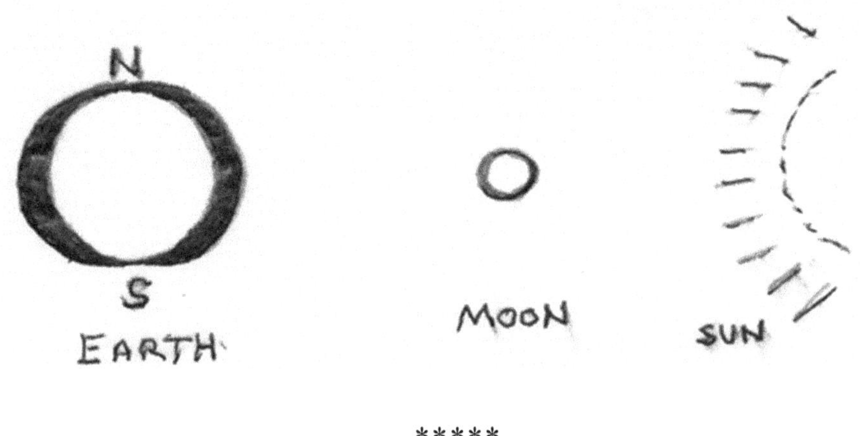

*****

So, in this observational guide, what would be a good starting point for our own planet?

A good place to start is your own area. What special features does the landscape have? Are there hills, mountains, lakes, rivers, rock formations there? How were they formed? What makes your area special? Are there different species of birds, animals or plants that thrive there? What does the Moon look like and when and where does it appear from your own particular viewpoint? Does the Sun appear from behind a tree or hill earlier or later at different times of the year?

On a global scale, great land changing events have taken place and continue to take place: earthquakes and volcanoes, great shifts in the tectonic plate boundaries. As will be explained in the next section, the Moon has an effect on coastal scenery through the relentless nature of the tides.

The Pyramids, Stonehenge, Great Wall of China, Machu Picchu, the scientific laboratories in Antarctica, aeroplanes and submarines stand testament to humankind's endeavours, tenacity, and ability to survive in all environments.

Earth has one moon, the details of which have been covered earlier. Suffice to say here that the Moon has a significant impact upon the Earth in respect of tides and that we are fortunate to inhabit Earth at a time to witness the Moon at a distance which gives rise to solar eclipses. The Moon is truly a great place to start an interest in astronomy and can be done during daytime as well as night-time. Observing the relative positions of the Sun and the Moon, particularly during an eclipse of either the Sun or the Moon, will really give you the feeling of being on a planet in a solar system full of moving objects. I use this in my talks about wellbeing as the vastness of space does tend to put things in perspective. Humans have long watched the sky, and many prehistoric monuments are aligned to certain stars as well as the Sun and Moon. There are an increasing number of Dark Sky Reserves which I urge you to visit and support. Encourage local councils to build new estates with low impact lighting i.e. the light shines down where it is needed with less impact on the night sky, thus enabling the stars to be seen and enjoyed. Ours is a unique planet in the vastness of the universe… until there is evidence of other life elsewhere. We are its custodians.

MARS

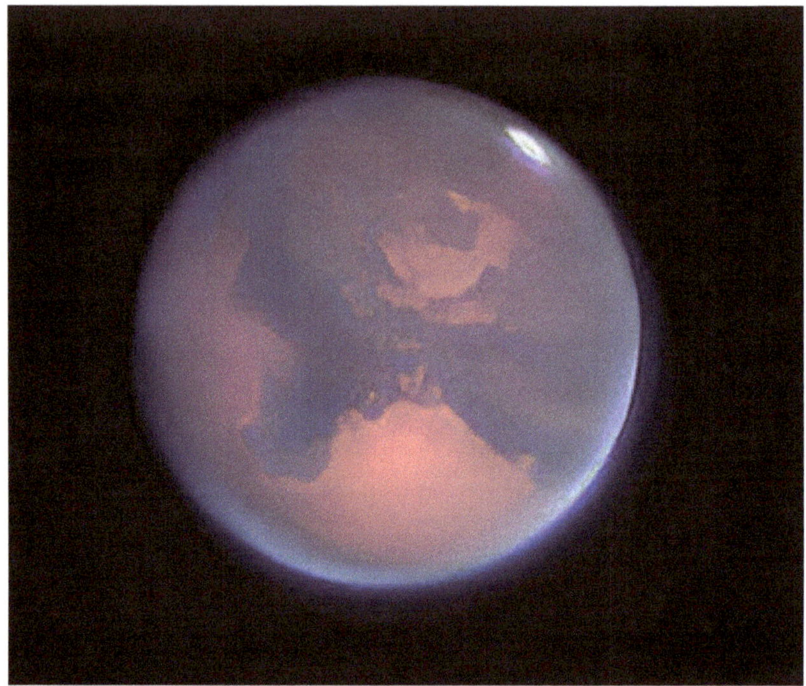

Credit: David Strange

Missions to Mars capture our imagination because it is the most likely destination for a manned spacecraft. Venus is nearer to us but is so inhospitable, landing there is out of the question. At least, for the time being!

So, what can we see through the telescope? If you are fortunate enough to have access to a telescope, it is possible to view the ice caps. I was fortunate enough to use Norman Lockyer's historic telescope at his Observatory in Sidmouth and found the ice caps all by myself. The effort involved was well worthwhile.

The planet appears as a red dot in the sky and the redness reminded the Ancient Greeks of blood and war, so they named it Ares. Later, the Romans changed the name to Mars – their god of war. The surface is rocky and littered with pebbles and rocks. It is very windy with tornadoes bigger than any on Earth. Visibility is poor with all the dust being blown around. Everywhere has a reddish tinge, due to iron rich content.

Some of the crevices are as deep as 8 km/5 miles and 644 km/400 miles wide. Each pole is covered by an ice cap. The crevices are evidence of water erosion. There are clouds on Mars, although they look similar to Earth clouds, these were formed from water and carbon dioxide in the Martian atmosphere and not from water on the surface which happens on Earth.

Olympus Mons is the largest volcano in our solar system. It is a shield volcano 15 miles high with a volume greater than any on Earth. Mount Everest is a bit less than 10 km/6 miles high (8839 metres/29,000 feet).

In an article in *Astronomy Now* magazine (May 2016) Dr Allan Chapman noted that the transit of Mercury on 3 June 2014 was the first such transit to be observed from a planet other than Earth. This was observed by the Mars rover, Curiosity.

Curiosity. Spirit. Opportunity. Perseverance. Such are the inspiring names of the missions to map and photograph and take samples of the red planet.

A tiny rover at the edge of a crater
Credit: Alison Young

The planet has two little moons encircling it, Phobos and Deimos, which would look like potatoes at the end of a football pitch!

Mars is half the size of Earth and takes twice as long to orbit the Sun. It comes closest to Earth every 780 days or so.

Between Mars and Jupiter is an area containing millions of small rocks ranging from a few feet/metres to about a kilometre across. The **Asteroid Belt** contains rocks some of which are larger than a kilometre across, but all are smaller than planets. Asteroid means starlike and these are small rocks that have never formed into planets or perhaps are pieces from a larger object which suffered a collision. Although far apart, these objects do present a danger to spacecraft travelling through on their way to the outer solar system.

JUPITER

This is a giant of a planet. Jupiter is a real favourite, offering plenty of things to see and to observe changes over time. The four moons are relatively easy to find in binoculars and it is fun to see which side of the planet they appear. It is one of the great gas giants, sitting far out in the solar system and dominating the night sky with its bright appearance. The planet radiates more heat than it receives from the Sun with a possibly liquid hydrogen core which makes it such a bright object. It is not a star though because it does not have the capacity for hydrogen fusion to take place as in stars.

The highlight of the planet itself is the **Great Red Spot.** This is a very long-lasting feature which comes and goes yet always does come back, its site marked by a large 'bay' area in the South Equatorial Belt. It is a gigantic whirling storm. It is large enough to contain two or three planets of Earth's diameter. Its colour is possibly due to phosphorus. The size varies over time, for example, in the late 19th century it was 40,000 km/28500 miles across, and more recently it was only 167,000 km/104,000 miles across. It is worthy of study over time to see for yourself the changes in size and colour. During the first dozen or so years of this century the red colour has intensified as the area has become smaller.

The bands that you see are different gases and all rotating at different speeds. The different elements in the atmosphere condense to give it this distinctive banded appearance. As different gases condense at different temperatures, different types of clouds form at various altitudes. Cooler air flows at a lower altitude and with the planet spinning as well; the larger circulating cell of air is split into many small ones of falling and rising air. These are thus seen on the surface as different bands of colour. The atmosphere is made up of 89% hydrogen, 11% helium and 1% other elements.

Jupiter is generally a very windy planet with winds that blow more than 400 mph. That is faster than hurricanes and tornadoes!

Jupiter is the fifth planet from the Sun. The distance between Earth and Jupiter varies between 4.2 AU to 6.2 AU because of their respective elliptical orbits. On average Jupiter is approximately 5 AU from the Sun, or 778 million km/484 million miles. At the equator its diameter is a massive 144,000 km/89,420 miles according to Patrick Moore's New Guide to the Planets.

Credit: Wikimedia Commons (cropped image)
https://commons.wikimedia.org/wiki/File:SolarSystem_OrdersOfMagnitude_Sun-Jupiter-Earth-Moon.jpg

It has a volume 1300 times that of Earth but would only weigh 318 times that of Earth due to the planet being constituted mainly of gases. Around 1,300 Earths would fit into Jupiter while about 1,000 Jupiters would equal the Sun's even greater volume.

The **Juno Mission** which launched on 5 August 2011, travelled 3 billion kilometres (2 billion miles) to arrive at Jupiter on 4 July 2016. Although due to end in 2021, the Mission continues to collect data from the planet and its moons and will finally end when the equipment no longer functions. At this stage the spacecraft will be programmed to descend deep into the atmosphere of Jupiter. Due to a technical issue the spacecraft only orbits the planet every 53 days instead of the planned 14 days, which has extended the length of the Mission. Data collected so far has revealed lower levels of ammonia and a different mix of gases than expected. It is also thought that Jupiter has a large core of heavy elements because of the gravitational pull on the spacecraft's orbit.

One of the exciting events witnessed by the spacecraft was the eruption of a white plume in the Jupiter's South Equatorial Belt. This plume came from an existing pale white spot.

This is a drawing of an observation made by John Meacham using a 12"
reflector at Bristol Astronomical Society Observatory. The markings in
the south polar belt are due to impacts from comet fragments.

The Moons of Jupiter

It is worth taking time to learn about Jupiter and its moons as this will give you a real sense of the vastness of space.

Stepping slightly beyond the remit of this book, there are little moons to observe around Jupiter through a telescope or binoculars. And, if you like moons, you will find plenty of interest as there are over 92 moons and still counting around this gas giant. The four largest moons are near Jupiter, although there are some smaller ones nearer still. These four moons were spotted by Galileo using an early telescope in 1610 and were the first moons to be seen orbiting another planet. Telescopes had evolved from the Dutch spectacle industry and were invaluable in enabling traders to compete to be the first to see merchant ships coming into port, in particular Venice, where the first person to be ready could take advantage of the trade opportunities. Galileo spotted an opportunity for their wider use.

These four Galilean moons are believed to have formed at the same time, and orbit in the same direction as Jupiter's spin, that is anticlockwise when viewed from above Jupiter's north pole. They are all tidally locked, which means they face the planet all the time. Io, Europa, and Ganymede have a particular relationship through their orbits. **For every orbit of Jupiter by Ganymede, Europa orbits twice and Io four times, i.e. a ratio of 1:2:4.** Every now and again, they are in a line and the gravitational effect of this line up means that their orbits remain elliptical. Callisto, however, is outside this effect. If you wish to learn more about this, the terms to research are 'orbital resonance' and 'Laplace resonance'.

Let's take a closer look at the four Galilean moons:

Io is larger and denser than the Moon and is 3,600 km/2,236 miles in diameter. At only around 422,000 km/262,000 miles out from Jupiter it is the closest of the Galilean moons. It takes just 42 hours to orbit the giant planet while also rotating once every 42 hours. There is a strong gravitational pull from Jupiter on one side with Europa on the other also exerting a strong pull. The effect of these gravitational forces on the surface is to keep the interior molten. The heat to keep the interior molten comes from the subsequent flexing and friction of the surface. The molten material erupts and constantly renews the surface. It is highly active with around 80 of the 300 discovered vents active. This young surface is yellow, orange, red and black from the sulphur that is emitted. There is a thin atmosphere of sulphur dioxide. Sometimes, the shadow of Io can be seen crossing Jupiter as the planet transits.

Europa, by contrast, is an ice-covered ball of rock, a bit smaller than our Moon. The dark lines may be the result of tectonic activity through volcanically heated water and ice. The water erupts through the surface and freezes instantly. This gives it a smooth surface, the smoothest of all the moons, of frozen water ice. There is a 100 km/ 62 miles deep ocean locked beneath the ice, containing more water than in all the oceans on Earth. The presence of water is suggestive of the possibility of life and is currently under investigation. Its orbit and rotation are just over 3 Earth days. It has a diameter of 3,000 km/1,800 miles and is just over 670,900 km/417,000 miles away from Jupiter.

Ganymede is the largest moon in the solar system, bigger than Mercury and three quarters the size of Mars. It is made up of rock and ice in a ratio of 60:40. The core is iron-rich with an upper mantle of ice and an icy crust. The darker areas of the crust are heavily cratered. There are layers of ice and water covering this moon. It is about 1.2 million km/665,000 miles from Jupiter. It takes 7 days to make its orbit and to rotate. Its diameter is 5,000 km/3000 miles.

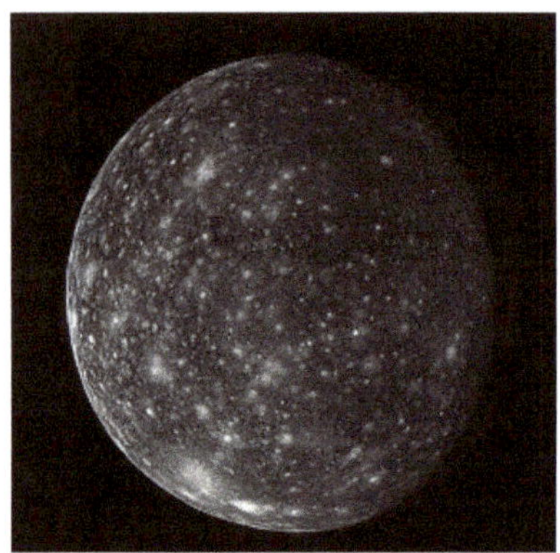

Callisto is the most distant and second largest of the Galilean moons at over *a million miles or 1.6 million km* from Jupiter. It is brighter than our Moon because its icy surface reflects sunlight. The moon is made up of a mixture of rock and ice, being slightly rockier at the centre. It takes nearly 17 days to orbit Jupiter and to rotate. It is nearly 4,800 km/2983 miles in diameter. The heavily cratered surface is an indication that this is an ancient and inert planet unlike the younger surface of Io.

So, when viewed through a telescope and 4 little dots are seen close to the big gas giant, it is worth bearing in mind the actual distances involved. Our Moon is just 360,000 km away from our planet, less than Io is to Jupiter. Below are the Galilean moons in a diagram:

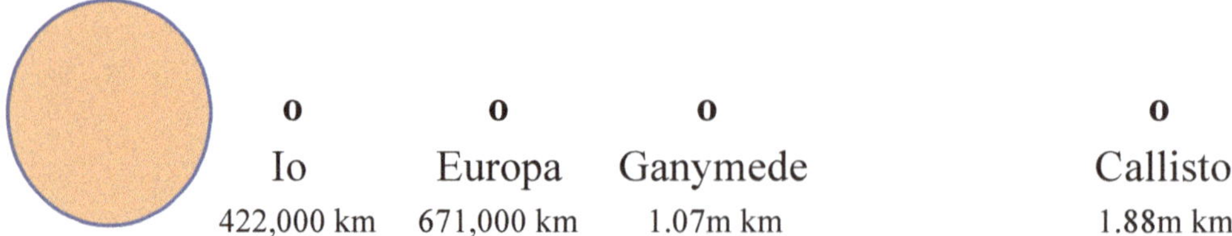

Jupiter with the four Galilean moons and their distance from Jupiter.

Using very average figures, assume Jupiter is 1,000,000 km diameter. On this basis, the distance from Jupiter for Io is 4 times the diameter, Europa is 6 times the diameter distant, Ganymede is 10 times the diameter distant, and Callisto is 18 times the diameter distant. In comparison, our Moon is 30 times the diameter distant from Earth (Earth's diameter being 12,700 km/7,918 miles). With Jupiter being ten times the size of Earth, its moons are surprisingly close to the large planet.

The tiny moons forming the inner group of Jupiter's moons are less than 200 km diameter and are within 200,000 km of the great planet. The outer moons are possibly fragments of other objects or captured asteroids. They orbit in the opposite direction to that of Jupiter. It is thanks to advances in astrophotography that have enabled many of these moons to be discovered in addition to data from mission flybys such as Voyager and from radio telescopes. The more distant moons are catalogued but not given names, such as S/2003 J2, one of the more distant of the outer moons. These moons are as far out as 30 million km/20 million miles and are around 2 km diameter (roughly a fifth of the distance from the Sun to Earth. Many of these moons orbit in the same direction as Jupiter's rotation. Following on from NASA's Juno mission, the European Space Agency launched its JUICE mission in April 2023 to further study the moons of Jupiter (**Ju**piter **Ic**y Moons **E**xplorer!). NASA's Europa Clipper Mission will study the moon of Europa in particular.

Jupiter and Io
Public Domain, https://commons.
wikimedia.org/w/index.php?curid=12683532

Earth and the Moon. The coin is a British 2p
Credit: Alison Young

SATURN

Image of Saturn taken at NLO

This planet takes just over 29 Earth years to orbit Sun. It is tilted on its axis so we can see above and below rings every so often. It is a fast-spinning planet, rotating on its axis every 10 hours as it orbits. This results in material moving to the equator, giving it its bulge. It is nearly 120,500 km/75,000 miles in diameter. The distance to the planet from the Sun is 1.43 billion kilometres/888 million miles. Like Jupiter, it too has a giant storm, the Great White Spot, first discovered by amateur astronomer Will Hay, a well-known comedic film actor in the 1930s. This also highlights the role of the amateur in astronomy due to the scale of the universe – the professional astronomers do not have time to invest in all the observation that is required.

Composition: It is made up of mainly gaseous (hydrogen and helium) material, with no discernible surface. It is so light it would float on an ocean (assuming one could be found large enough!). It is the least dense of all the planets. Although its mass is 95 times that of Earth, due to its immense size, 764 Earths could fit inside it.

Ring System: Ranges from 5 to 30 km/3 to 19 miles deep. Tidal forces just inside the ring area are so strong any moon would be torn apart.

Cassini Division: The Cassini Division is a region 4,800 km (3,000 miles) in width between Saturn's A Ring and B Ring. Apparently empty as seen from Earth but contains thousands of ringlets. It is helpful to think of the composition of the rings appearing like the ice on an arctic ocean.

Around Saturn are 60+ moons, and small (shepherd) moons exist within the rings. These influence the movement of the rings in very small changes.

Cassini Mission: Launched on 15 October 1997 and continued until September 2017, the Cassini spacecraft with its Huygens probe entered orbit around Saturn on 30 June 2004. Giovanni Cassini discovered the divisions in the ring system in 1675 and Christiaan Huygens discovered Titan in 1655 – the time of Tudors and Stuarts. It took 60-80 minutes for signals to travel each way from the spacecraft to Earth. It was a successful mission with more data from the different moons being gathered than originally anticipated.

The spacecraft Cassini released the Huygens probe to parachute through Titan's atmosphere. As it descended, the probe recorded rivers and deltas carved by methane rain. Storms were found in Saturn's atmosphere. The aim of sending the probe deep into the gasses of Saturn for the final stage of the mission was to try to get more data and then to not have another piece of space junk around which may contaminate the pristine moons such as Enceladus which may contain elements for life.

This may be the last mission with so many scientific instruments on board. The current design tends to be for single focus mission which are not as versatile, and, as such, constrains research. Cassini was able to be redirected and could send back all sorts of data.

Titan

URANUS

Observe the bands going around from top to bottom, rather than the more normal horizontal banding. (NASA Image)

This planet was discovered in 1781 by William Herschel, by chance whilst looking for comets. Initially he named it George, after George III. However, in keeping with the standard naming system for planets it was officially called Uranus. Uranus being the father of Saturn... who in turn is the father of Jupiter in Greek mythology. It is the eighth planet from the Sun.

On its side... In comparison with the other planets, it is sideways on. Long ago, a massive object must have crashed into Uranus, changing its orbital direction, or perhaps a large moon was pulled away when the solar system was still forming, with its gravitational pull causing Uranus to tilt on its side.

It is a gas planet with a liquid centre. The gas is made up of hydrogen and helium. A small amount of methane gives it the colour blue. At -224°C it is the coldest planet in the solar system. Herschel thought he had seen rings but this was discounted at the time. However, a set of rings were indeed discovered in

1977 and a second set was seen by Hubble Space Telescope in 2003. Rings may be made up of dust particles and large rocks.

There are a great many little moons, around 27 altogether. The 5 large ones: Titania, Oberon, Umbriel, Ariel and Miranda are named after characters from Shakespeare.

William Herschel was a kind man, taking care of his sister (Caroline) who was unable to marry due to her skin being pockmarked from illness. Caroline became an astronomer in her own right and is a subject in herself... for another time. They lived in Bath where Herschel made lenses. There were workshops in every room! William was undertaking a full review of the sky when he spotted a new object which turned out to be another planet; initially he had thought it was a comet. His discovery was verified by another astronomer and its correct orbit was worked out by mathematicians – one of whom was executed during the French Revolution. The planet had been observed before but taken to be another star. It can be seen with the naked eye as a distant star. Through a telescope it appears as a pale bluish disc.

NEPTUNE

You really begin to get a feel for the vastness of space when you consider that Neptune is about as far from Uranus as Uranus is from the Sun. In other words, Neptune is 1,626 million km/1,010.4 million miles from Uranus while Uranus is 2,871 million km/1,784 million miles from the Sun. One of its moons, Triton, may even be a captured Kuiper Belt object from the outer regions of our solar system.

This planet is a ball of predominantly methane gas which gives it its blue colour. Methane reflects blue and absorbs red light. It is very windy although the cause for this is as yet unknown. The speed of the wind is around 2092 km/1300 miles an hour.

In 2020, the planet had just gone around the Sun once since its discovery in 1846. It takes 165 years to orbit the Sun although it rotates rapidly in just 16 hours. It has a similar tilt to that of Earth, with seasons lasting 40 years. Sunlight takes 4 hours to reach Neptune.

The discovery of Neptune is an interesting story. Scientists knew that an unknown body was influencing the orbit of Uranus. Its existence was then predicted by two people independently; English mathematician John Couch Adam in 1843, and French scientist Le Verrier in 1846. Their data in turn was used by Berlin astronomer, Johann Galle and his assistant, to actually observe it.

Neptune has 14 moons, the largest of which is Triton. Triton is slightly smaller than our moon, making the other moons very small. Triton moves around Neptune in a retrograde motion and eventually the effects of the tidal interaction will slow its orbit, or it will fall into Neptune. The moon Nereid was discovered by Gerard Kuiper in 1949. In 1989, 6 more little moons were found by the Voyager 2 spacecraft. Since then, five more small moons have been found. When Voyager 2 passed by, uneven rings were observed, which were possible from a collapsed moon. One of these rings was named after Adam. The latest images from the James Webb telescope clearly show the delicate rings around the planet.

And last, but not least…

PLUTO

Image credit: NASA/Johns Hopkins University Applied Physics Laboratory/Southwest Research Institute/Alex Parker - http://pluto.jhuapl.edu/Galleries/Featured-Images/image.php?page=1&gallery_id=2&image_id=543

Having grown up with knowing Pluto as a planet, I have a soft spot for it now that it has been demoted to a **dwarf planet**. It will always be a planet to me although it is no longer officially one of the main planets in our solar system. Regrettably, I no longer have a poster of the solar system which included Pluto. This was displayed on the wall of my childhood bedroom in the way other people have posters of pop stars.

What is the difference between a planet and a dwarf planet? To be a planet, an object needs to have cleared the area around them. It was controversially demoted to dwarf planet in 2006. Dwarf planets are **small**. Part of a larger system of smaller objects, rather than being on its own in orbit like the main planets. Lots of them. Still being discovered. Pluto has 5 moons including Charon. It has an odd relationship with Neptune as their paths cross occasionally due to the idiosyncratic nature of their orbits, with Pluto then nearer to the Sun than Neptune. While Neptune is 30 AU from the Sun, Pluto varies between 30 and 49 AU. When Pluto is 'closest' to the Sun, the temp is 10°C higher and causes changes to the surface and atmosphere (a process called sublimation). It takes 248 years for this tiny planet, around three quarters of the size of our Moon, to orbit the Sun.

The planet was discovered between 23 and 29 January 1930 at Lowell Observatory Arizona. This was set up by Percival Lowell (an eminent Victorian businessman, mathematician and astronomer born in 1855) who predicted its existence but had not found it. The planet was discovered 14 years after his death in 1916, aged just 61. It is the oldest observatory in the USA.

A naming competition was held which was one by an 11-year-old schoolgirl, Venetia Burney (1918-2009), who named it Pluto, god of the underworld.

The discoverer of Pluto was Clyde Tombaugh who was born on 4 February 1906, he built his own telescope at the age of 22 and was around 30 years old when he gained his astronomy degrees. He was subsequently offered a job at Lowell Observatory to seek the planet that would be called Pluto. He died on 17 January 1997.

*****

If Pluto is not one of the main solar system planets, then where is it? It is within the area now known as the Kuiper Belt where there are many more similar objects to it. Pluto was the first Kuiper Belt object identified in 1930, but the next one was not found until 1992, which was Eris. Pluto's companion, Charon, was discovered 48 years after Pluto, in 1978.

According to *Astronomy Now* (June 2016), the effect of the solar wind upon Pluto means it conforms neither to a planet nor a comet. It is indeed neither but is now classed as a minor planet. Pluto is made up of ice and rock and is just 2000 km/1242 miles in diameter. The surface is moving at the relatively speedy rate of 2cm/0.75 inch per year; geologically speaking this is indeed rather fast, with glaciers 'flowing' down. There is a large impact crater area, Sputnik Planum, on the far side of Pluto, opposite Charon.

The New Horizons spacecraft is about the size of a small grand piano. The mission lasted for 6 months around Pluto, setting off on 19 January 2006, to fly-by Pluto on 14 July 2015. It is powered, appropriately by the radioactive decay of plutonium.

NASA image

Finally, to round off this section about the planets, below is the awesome image of our *Pale Blue Dot* (the tiny bright dot in that streak of sunlight) taken by Voyager 1 on 14 February 1990 as it looked back to our home planet from the outer reaches of the solar system:

A fun activity which highlights the vast distances between the planets is to take one sheet of paper to represent one Astronomical Unit (1 AU) which is the measurement used for the distance of the Earth from the Sun. On that sheet of paper, you need to mark on Mercury and Venus as they fall within 1 AU from the Sun. Then, using other pieces of paper, place the paper at the appropriate intervals from the Sun. 40 sheets of paper will take you out to Pluto.

PLANET	AU FROM SUN	DISTANCE FROM SUN
MERCURY	0.4	57 million km
VENUS	0.7	108 million km
EARTH	1.0	150 million km
MARS	1.5	228 million km
ASTEROID BELT	3.0	400 million km
JUPITER	5.0	779 million km
SATURN	9.5	1.43 billion km
URANUS	19	2.88 billion km
NEPTUNE	30	4.50 billion km
PLUTO	40	5.91 billion km

PLANETS AT A GLANCE

MERCURY	
How far from the Sun?	57.9 million km/36 million miles
Diameter	4,875 km/3,029 miles
Mass (Earth = 1)	0.055
Volume (Earth = 1)	0.056
Time taken to orbit the Sun	88 Earth days
And…	It is very, very hot by day (430°C) and very, very cold by night (-180°C)

VENUS	
How far from the Sun?	108.2 million km/67.2 million miles
Diameter	12,104 km/7,521 miles
Mass (Earth = 1)	0.82
Volume (Earth = 1)	0.86
Time taken to orbit the Sun	224.7 Earth days
And…	Nicknamed the 'Coca Cola' planet because of all the carbon dioxide in the atmosphere.

EARTH	
How far from the Sun?	149 million km/93 million miles
Diameter	12,742 km/7,917 miles
Mass (Earth = 1)	1
Volume (Earth = 1)	1
Time taken to orbit the Sun	365 days, 5 hours, 59 minutes and 16 seconds
And…	Named the 'Blue Planet' as it appeared blue to astronauts from their vantage point of space.

MARS	
How far from the Sun?	227.9 million km/141.6 million miles
Diameter	6,780 km/4,213 miles
Mass (Earth = 1)	0.11
Volume (Earth = 1)	0.15
Time taken to orbit the Sun	687 Earth days
And…	There is ice at the poles.

JUPITER	
How far from the Sun?	778.3 million km/483.6 million miles
Diameter	142,984 km/88,846 miles
Mass (Earth = 1)	318
Volume (Earth = 1)	1,321
Time taken to orbit the Sun	11.86 Earth years
And…	Galileo discovered 4 of the planet's inner moons in 1610.

SATURN	
How far from the Sun?	1.43 billion km/888 million miles
Diameter	120,536 km/74,898 miles
Mass (Earth = 1)	95
Volume (Earth = 1)	764
Time taken to orbit the Sun	29 Earth years
And…	Even in the seemingly dark gaps between the major rings, there are thin rings containing tiny particles of rock and ice. It only takes 10 hours to rotate.

URANUS	
How far from the Sun?	2.87 billion km/1.78 billion miles
Diameter	51,118 km/31,763 miles
Mass (Earth = 1)	14.5
Volume (Earth = 1)	63.1
Time taken to orbit the Sun	84 Earth years
And…	Another fast-spinning planet, it takes just over 17 hours to rotate.

NEPTUNE	
How far from the Sun?	4.5 billion km/2.8 billion miles
Diameter	49,532 km/30,760 miles
Mass (Earth = 1)	17.1
Volume (Earth = 1)	57.74
Time taken to orbit the Sun	164.9 Earth years
And…	This has a bulgy middle due to its fast spin, taking just over 16 hours to rotate.

PART 4

A SELECTION OF TOPICS

Stars

Sun, The

Magnitude and Norman Pogson

Comets

Meteor Showers

Norman Lockyer and the NLO

Observing Equipment

How to use a planisphere

Observing Log (blank)

STARS

This is a huge topic, as vast as the universe. Therefore, this is an overview to help you understand what stars are. There is a wealth of good astronomy books available if you wish to take the subject further.

So, what is a star? It is a ball of gas with nuclear fusion taking place within it. Stars twinkle due to atmospheric disturbance; whereas planets have a broad disc and are less prone to disturbance and therefore do not twinkle. This makes it easy when you look up to see whether what you are looking at is a star or a planet.

Stars shine by fusing hydrogen atoms to form helium. Eventually they exhaust the primary fuel and begin dying. Brighter stars burn out more quickly because they use much higher temperatures and pressures at their core, using up the supply of hydrogen quicker. Interestingly, on Earth there are various uses of helium, for example, it is used in cooling superconducting magnets in MRI scanners.

The following diagram gives some idea of the scale of the distance to the nearest star by comparing how long it takes light to reach Earth with how long it takes light from the Earth to reach the nearest star.

Sun .... Earth ....Proxima Centauri

< 2.5cm = 8½ light *minutes* > <.....................6 km = 4 light *years* ....>

(Proxima is part of Alpha Centauri star system)

Stars are formed within giant dusty gas clouds. Deep down in a cloud a clump grows big enough that the gravity pulls more gas into the cloud. The clump of gas collapses and heats up to form a star. The life cycle of stars is complex and is illustrated on a Hertzsprung-Russell diagram. The diagram illustrates how stars fall into groups; our Sun is in the main sequence which means that such stars are at their most stable before evolving into red giants or supergiants.

However, not all stars are born, live and die. It is much more complex than that. Pulsars (pulsating radio stars) are what remains of massive stars. These generate more nuclear fusion as they rotate, thus constantly renewing their source of energy.

The colour of a star gives a rough guide to its age or stage in its life cycle. Young, hot stars are blue or white, which are hotter than our own yellow star. The older stars are red giants, stars which have exhausted the supply of hydrogen in their cores and switched to thermonuclear fusion of hydrogen atoms to form helium in a shell surrounding the core. They have radii tens to hundreds of items larger than that of the Sun. However, their outer envelope is lower in temperature, giving them an orange hue. Red giants are many times more luminous than the Sun because of their large size.

An example of a red giant is the Garnet Star. Mu Cephei, to give it its proper title, is nearing death. It has begun to fuse helium into carbon, whereas a main sequence star fuses hydrogen into helium. When a supergiant star has converted elements in its core to iron, the core collapses to produce a supernova and the star is destroyed, leaving behind a vast gaseous cloud and a small, dense remnant. For a star as massive as Mu Cephei the remnant is likely to be a black hole.

Greek Alphabet

The letters of the Greek alphabet are used to denote the stars in a constellation in order of brightness. The main star in a constellation is *a* or alpha, going down through the alphabet in decreasing magnitude.

THE GREEK ALPHABET					
Alpha	A α	Iota	Ι ι	Rho	P ρ
Beta	B β	Kappa	K κ	Sigma	Σ σ
Gamma	Γ γ	Lambda	Λ λ	Tau	T τ
Delta	Δ δ	Mu	M μ	Upsilon	Y υ
Epsilon	E ε	Nu	N ν	Phi	Φ φ
Zeta	Z ζ	Xi	Ξ ξ	Chi	X χ
Eta	H η	Omicron	O o	Psi	Ψ ψ
Theta	Θ θ	Pi	Π π	Omega	Ω ω

Standard Candles are stars whose known brightness is used to determine the brightness of other stars. They are stars of known brightness which varies over time. Their characteristics have been the subject of much study and thus can be compared to new objects. An example of such a star is Mu Cephei or the Garnet Star. The time they take to change from bright to dim and back again is very regular and related to their absolute magnitude or how bright they actually are. They pulsate by the expansion and contraction caused by a layer of ionized hydrogen and helium in the deep levels of the star. (*Ionization is the physical process of giving or taking away electrons from an atom.*) This makes the layer opaque or impenetrable to light. The star expands against this layer until its opacity drops enabling energy to escape and thus the star contracts. And the process repeats. Their luminosity can vary as much as two magnitudes and their diameter as much as 10%. The temperature can also vary as much as 1000°C.

The relationship between a star's spectrum and its luminosity was studied by H. N. Russell of the United States and E. Hertzsprung of Denmark. The stars are classified into five main luminosity classes:

I Supergiants which are very massive and luminous stars near the end of their lives. These stars are very rare, one in a million stars is a supergiant. The nearest supergiant star is Canopus, 310 light years away. Some other examples are Betelgeuse, Antares and Rigel.

II Bright Giants are stars which have a luminosity between the giant and supergiant stars. Some examples are Sargas and Alphard.

III Normal Giants are mainly low-mass stars at the end of their lives that have swelled to become a giant star. This category also includes some high mass stars evolving on their way to supergiant status. Some examples are Arcturus and Aldebaran.

IV Subgiants are stars which have begun evolving to giant or supergiant status. Procyon is entering this category.

V Dwarfs are all normal hydrogen-burning stars. Stars spend most of their lives in this category before evolving up the scale. Class O and B stars in this category are actually very bright and luminous and generally brighter than most Giant stars. Some examples are the Sun, Sirius, and Vega.

THE SUN

Halo round the Sun May 2020
Never look at the Sun directly -
I was lucky to spot a well-placed streetlamp, enabling the halo to be visible safely

The Sun is an unusual star, as most stars have companions, and it does not. It is about 4.6 billion years old and is 750 times the mass of all the solar system's planets together. It takes one million years for photons (particles of light) to travel through the inner radiative zone before escaping through the photosphere as heat and light. The light from the Sun takes 8 light minutes to reach Earth.

The Sun's Interior

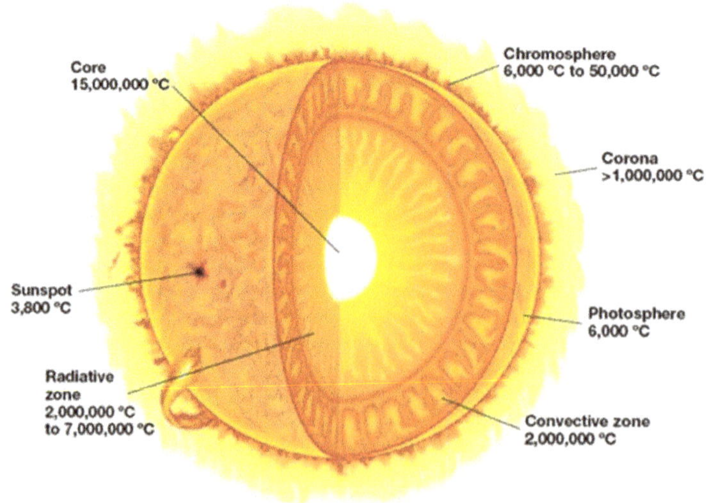

The **photosphere** (4500°C) is the visible surface of the Sun that we are most familiar with. Since the Sun is a ball of gas, this is not a solid surface but is actually a layer about 100 km thick (very, very, thin compared to the 700,000 km radius of the Sun). Toward the centre of the disk of the Sun are hotter and brighter regions. When we look at the limb, or edge, of the solar disk we see light that has taken a slanting path through this layer, and we only see through the upper, cooler and dimmer regions. This explains the 'limb darkening' that appears as a darkening of the solar disk near the limb.

The **Chromosphere** (literally, 'sphere of colour') at 20,000°C is the second of the three main layers in the Sun's atmosphere and is roughly 2,000 km deep. It sits just above the photosphere. It was found that the temperature of this layer of the solar atmosphere increases with increasing height in the chromosphere itself. The temperature at the top of **photosphere** is only about 4,400°K, while at the top of **chromosphere**, some 2,000 km higher, it reaches 25,000°K.

The **Corona** is the Sun's outer atmosphere of low-density gas with neutral impact on atoms. It is visible during total eclipses of the Sun as a pearly white crown surrounding the Sun. The corona displays a variety of features including streamers, plumes, and loops. These features change from eclipse to eclipse and the overall shape of the corona changes with the sunspot cycle. However, during the few fleeting minutes of totality few, if any, changes are seen in these coronal features. The flares, or prominences, coming from the Sun are cooler eruptions.

* NEVER LOOK AT THE SUN DIRECTLY *

Sunspots are exciting features to observe using a telescope with a piece of paper under the eye piece upon which to project the Sun.

The above photo was taken by me at the Norman Lockyer Observatory, where the sunspots were projected onto paper using the Lockyer telescope. Each spot is about the size of Earth, and they move in pairs across the surface of the Sun. The spots are darker than surrounding area because they are cooler. Interestingly, if you were to put a sunspot in the night sky, it would glow brighter than the Full Moon with a crimson-orange colour. As an illustration, a very hot pizza oven is around 300°C and these cooler spots are in the region of 3000°C–4500°C, about 2000 degrees cooler than the surrounding area. They are temporary features with an approximate 11-year cycle of activity. They appear mainly near the Sun's equator and move outwards in pairs expanding and contracting as they move. They can move at speeds of around a few hundred metres per second.

Solar flares can be seen through a special filter. These are when the Sun sends out magnetic material (coronal mass ejection) reaching Earth, sometimes causing mayhem particularly in northern USA and Canada.

* NEVER LOOK AT THE SUN DIRECTLY *

*****

The ecliptic traces out the apparent path the Sun takes during the year through the sky from our perspective on Earth. The name comes from the fact that eclipses can only occur along it. A lunar eclipse happens when the moon passes through Earth's shadow, when it is directly opposite the Sun on the sky. During a solar eclipse the moon passes between Earth and the Sun momentarily blocking out its light and warmth. Though the moon circles Earth roughly once a month, eclipses don't happen nearly that frequently because the moon's orbit is slightly tilted relative to that of our planet. Our moon spends most of its time either above or below the plane of Earth's orbit and therefore is usually not nicely aligned with us and the Sun.

As the other seven planets orbit in approximately the same plane as Earth, the ecliptic is also a guide to where you'll see the planets in the sky. Alternatively, to look at it another way, the planets allow you to actually trace out the ecliptic yourself.

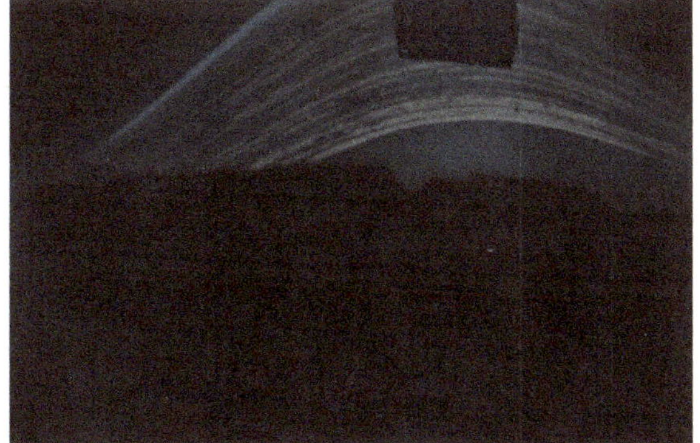

This image was produced using a pin-hole camera over the course of three months (Jan-Mar). The white lines are the path the Sun took during that time, gradually becoming higher in the sky as the Earth tilted towards it.

This is a painting made from a sketch at the time of a solar eclipse on 29 March 2006 in Antalya, Turkey by John Meacham with the Cardiff Astronomical Society.

MAGNITUDE

Magnitude is a measure of the varying brightness of stars when viewed from Earth. The scale is based on Vega. There are two types of magnitude: **apparent magnitude** which is the brightness of a celestial object as observed from Earth. **Absolute magnitude** relates to the brightness of an object corrected as to how bright it would appear if all the stars were the same distance from Earth.

Vega: This star is 25 light years away and is two and a half times the size of the Sun. It is familiar as one of the stars of the Summer Triangle. The Greek astronomer, Hipparchus, labelled the first stars to appear after sunset as alpha because they were the brightest (apparent) down to sixth magnitude for those stars visible only in darkness. Later, Norman Pogson developed this idea and introduced a scale of magnitude, the Pogson Scale.

The brighter an object appears, the lower the value of its magnitude, which means the brightest objects have negative values. The Sun has an apparent magnitude of −27, the Full Moon −13, the brightest planet Venus measures −5, and Sirius, the brightest visible star in the night sky, is at −1.5.

The Pogson Scale

Each increase of magnitude is two and a half times greater than the next value.

The size of dots on star charts represents visually the brightness of stars as they appear to astronomers on Earth. For variable stars, the dots have a smaller dot in the middle.

Logarithmic scale: difference ratio of 2.5 i.e. difference of 5 magnitudes equates to a difference of 100 in the brightness of the stars concerned. A difference of 1 magnitude is a ratio of 2.5, the fifth root of 100.

The man behind the scale: Norman Pogson

The story of the man behind the Pogson scale is a tragic one. He died in June 1891 after a lifetime of working and trying hard with his greatest idea going unnoticed at the time. With the particularly rigid class system, he was unable to rise above his humble family roots. At the age of 16 he ran away from Nottinghamshire to London teaching arithmetic to support himself. He did become assistant to an astronomer and the following year got a post at Oxford's Radcliffe Observatory. He was not supported and was alienated from his colleagues because he had not gone to university. With 11 children to support he accepted the post of director of a private observatory complete with a house as well as a good salary. He unsuccessfully applied for a post at Oxford. It was a case of who you know, and the Oxford post went to their own favoured candidate. In dire financial straits he applied for the vacant post of Government Astronomer in Madras and the family sailed there in 1861. Here he was well supported by the local astronomer and worked on a star atlas of the southern sky and variable stars,

Image credit: Popular Astronomy 1913

assisted by his daughter, Isis. One of his last discoveries was an asteroid which he named after one of his children, Vera. Vera died in infancy and was one of three from his second marriage. His first wife having died in India. And some of his children. This is particularly sad as they sailed to India with the knowledge that this outcome was most likely. They really must have been desperate.

Pogson made a footnote in his research paperwork suggesting the system of star magnitudes, where they increase by a factor of 100. This is now the international system of star magnitudes, the Pogson Scale. Sadly, the footnote was only fully understood after his death so he was not to know his name would live on.

COMETS

Comets are exciting, memorable events as they are sometimes seen chugging across the sky. They are easy to spot, with long bright tails of light. Indeed, the name comet comes from the Greek kometes meaning long haired. I am thinking in particular of a couple that I have seen, Hale Bopp and the Great Comet of 2007.

These two could be seen night after night making their way across the sky. A well-known example of a comet is that of Halley's, which is depicted in the Bayeux Tapestry. As is the tradition with comet discoveries, Edmund Halley gave his name to the comet he discovered and which has a 75-year orbital period, which frequency has helped this one become one of the more familiar and well known. But what are comets and where do they come from?

They originate from two main areas, the Asteroid Belt, and the Oort Cloud region. From the Asteroid Belt, between Mars and Jupiter, originate those with a short orbital period, while those with a long orbital period come from the far reaches of the outer regions of our solar system. The sphere-shaped edge of the solar system known as the Oort Cloud is situated around 50 times further out than the Kuiper Belt. The Kuiper Belt is just beyond the orbit of Neptune and includes Pluto and other dwarf planets. And long does indeed mean long, with Hale Bopp having an orbital cycle of some 2500 years.

Recent research has shown comets to be essentially snowy dirtballs. In a talk at Exeter, Mark McCreaghan of European Space Agency suggested that comets are left over from the formation of the universe which means they hold very ancient material. For some reason they missed being part of a planet and are too small to be classed as planets on their own. Some become comets due to collision with other objects which then take their orbit into the solar system. Those from the Asteroid Belt are quite likely to contain pieces left over from planet formation, while those from the Oort Cloud are more likely to contain general debris.

An image taken on 24 July 2023 of the so-called 'horned' comet in a gap in the clouds with RC10 + ASI533MC. It was discovered in outburst having brightened 5 magnitudes over the last few days. Courtesy David Strange, NLO.

Taking a closer look at the structure of a comet, the coma is a dusty atmosphere surrounding the nucleus or core. The strong solar winds blow the dusty material into space, forming the two jets of gas and dust which always point away from the Sun. The speed of comets varies roughly between 10 km and 70 km.

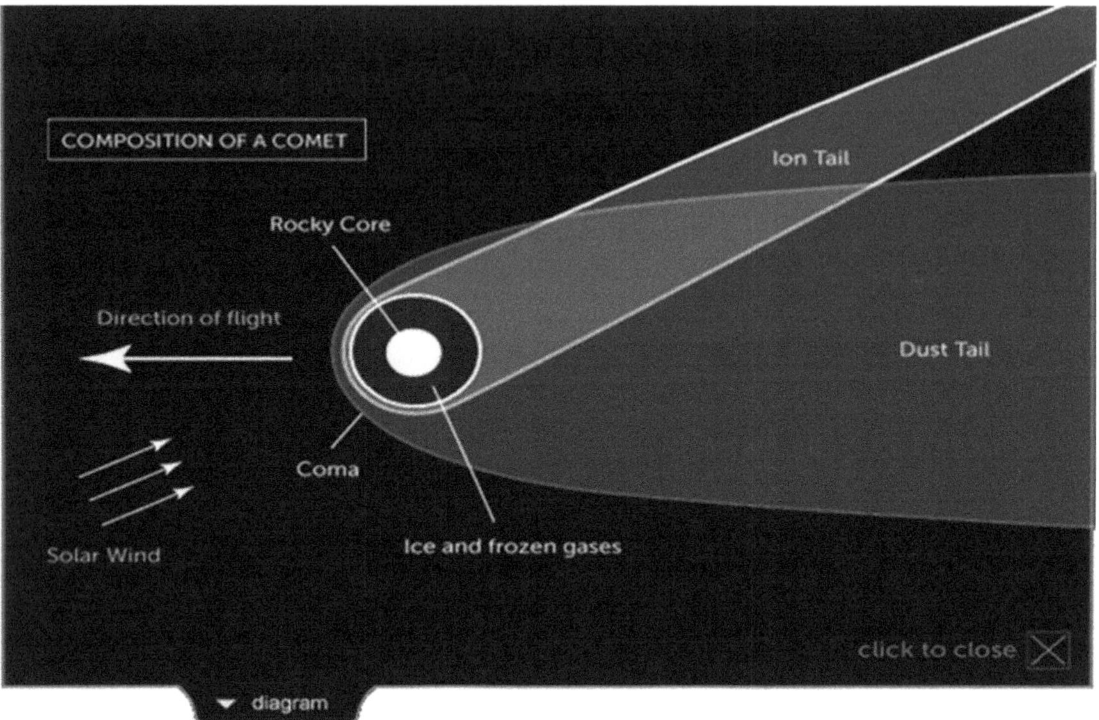

The intense heat from the Sun causes the harder crust to soften and for particles to fall away, forming a tail. Their insides are of softer material including oxygen and ice. Out beyond Jupiter, the heat has little or no power and the comets are cold and lifeless, continuing to hurtle through space on their long trajectories.

The Rosetta Mission

This is a good starting point for learning more about comets. And, yes, it included a cuddly toy!

Thanks to my personal circumstances at the time, I was able to indulge in wall-to-wall viewing of the final throes of thc Rosetta Mission to Comet 67P. This Mission enabled humans to land an object on a fast-moving comet for the first time. This is an incredible achievement and the whole process began back in the 1990s following research into Halley's Comet when it passed by in 1986. Rosetta was launched on 2 March 2004 and ended on 29 September 2016. The staff were in a sombre mood as their waited for the transmitting signal from Rosetta to end. Normally it is the reverse, and they cheer when contact is established. This was very different. This was the end, the absolute end of a mission. It was programmed to switch off upon impact.

The lander Philae had only been moderately successful as it had bounced on landing into an overhang rock. It landed on 12 November 2014 taking a panoramic view of the surface and somc data about molecules. The descent lasted 7 hours. And then went to sleep for a long while until the news was broadcast that Philae had woken up (summer 2015)!

Images were taken by Rosetta as it descended to the comet's surface. It landed on the comet at a walking pace. This was an amazing achievement as it was never planned that Rosetta would land on the comet too.

Comet 67P contained components of DNA – the building blocks of life. The comet was formed from two lumps coming together. The pits seen are the result of massive heating and cooling and offer an insight into the heart of the comet.

Comet 67P Churyumov/Gerisamenko is named after the two people who discovered it. Klim Ivanovich Churyumov died in 2016, aged 79 years, fortunately *after* the successful mission. Svetlana Gerisamenko photographed the comet in 1969.

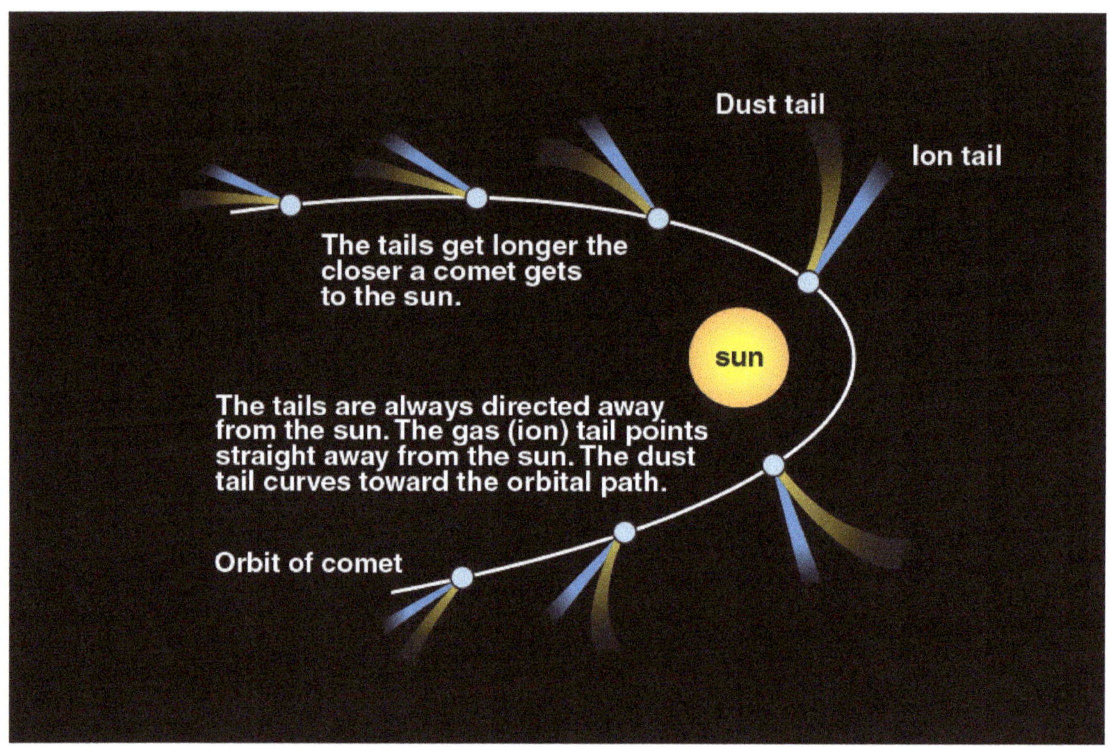

Credit for Comet around the Sun diagram: By NASA Space Place - Own work, CC BY-SA 4.0, https://commons.wikimedia.org/w/index.php?curid=54176004

*****

How do you know it's not a comet? It is not a comet if an object is listed in the Messier Catalogue… Charles Messier was an avid comet hunter and to make his life easier, he compiled a catalogue of objects he had initially thought were comets but were not, in order to help other comet hunters. What he achieved was a handy catalogue of galaxies, globular clusters and nebulae which some observers use to enjoy an evening trying to spot them all – a Messier Marathon. It is a list of wonderful objects to enjoy and includes coordinates for them all. So, if you can work out the coordinates of an object you can easily check what it is or is not.

METEOR SHOWERS

This is a meteor from the Orionid shower

If you enjoy catching a glimpse on an elusive shooting star, then you will certainly enjoy a more targeted approach to meteor watching (see table below). Meteors (the correct name for shooting stars) are the result of debris from what remains of a tail of a long-gone comet through which the Earth passes periodically. Some meteors are from other sources such as manmade space debris or just debris.

Those fortunate enough to be on holiday during August, can treat themselves to an interrupted night's sleep by setting your alarm to wake you during the night. At this time of year, the Perseid meteor shower is an enjoyable treat... given a clear night of course! There are other meteor showers throughout the year... but at less warmer times, so make the most of this particular shower.

Meteors are any material (some from comets, some manmade, some of unknown origin) which burn up on entry through the Earth's atmosphere, the friction between them and the atmosphere causing the brightness. They appear as a bright flash or meteor. If they hit the ground, they become *meteorites*.

Meteor showers originate from the constellations bearing their name. This is known as the radiant or the area of sky from which they originate. Meteors or meteoroids shoot across the sky and when they hit the Earth's atmosphere (the protective bubble around us) they burn up and fall to the Earth as dust. Sometimes larger pieces of the rock survive and land as meteorites. Meteors can be glimpsed all year round but with known showers offering a more reliable opportunity of seeing one or more as listed in the table below. They are more visible overhead where the 'seeing' is less affected by atmospheric disturbance. If you look closely, you may see that they are different colours. This is due to the chemical composition of a particular meteor. As they burn up when entering the atmosphere different colours are produced according to their chemical makeup, for example, one made mainly of calcium will emit a purple or violet colour while magnesium will have a greenish colour.

Regular Meteor Showers

These are visible either side of the main date in those quiet hours before dawn… (it is advisable to check the press for up-to-date information and weather before embarking on a session. You may come across the Zenithal Hourly Rate, this is just the term given to the number of meteors likely to be seen directly overhead). Did I mention it might be a good idea to use a sun lounger? You will be looking up at a wide area of the night sky so lying down on something dry and comfortable is a good idea.

28 December - 12 January	Quadrantids	Near the end of the Plough (their original constellation of Quadrans is no more)/Asteroid 2003 EH1).
16 April - 25 April	Lyrids	Lyra/comet C/1861 G1 Thatcher (named after A. E. Thatcher who discovered it in 1861).
19 April - 26 May	Eta Aquariids	Aquarius/Comet Halley (very low on the horizon, emanating from the star Eta aquaria).
22 May - 2 July	Arietids	Aries/unconfirmed origin.

12 July - 23 August	South Delta Aquariids	Aquarius/Comet 96P Machholz possibly.
13 July - 26 August	Perseids	Perseus/Comet Swift-Tuttle.
4 October - 14 November	Orionids	Orion/Comet Halley.
5 - 30 November	Leonids	Leo/Comet Temple-Tuttle
4 - 16 December	Geminids	Gemini/Asteroid 3200 Phaethon

Meteorites can have a huge impact on Earth as was the case for the dinosaurs when a massive meteorite landed on the Yucatan peninsula. This caused the climate to change with the knock-on effect of a decline in the food supply for the dinosaurs. The Tunguska area in Russia is well documented for its trees that have been knocked over by the explosion just above them of a massive meteorite in 1908. Other meteorites of note include one seen to land in a garden in England, residents of a house in Winchcombe, Gloucestershire saw black marks on their driveway the morning after a fireball (a large meteor) was seen in the skies on 1 March 2021 and subsequently collected several bits of space rock, now in the Geological Museum in London, and in Chelyabinsk, Russia in 2013 a big fireball was filmed by drivers using their dashcams, which has since become a phenomenon in Russia with subsequent meteors being filmed in that way.

NORMAN LOCKYER
AND THE
NORMAN LOCKYER OBSERVATORY

I have mentioned Norman Lockyer and my local observatory, the Norman Lockyer Observatory in this book and would like to draw the book to a close with some information about them.

Sir Norman's notable achievements are the identification of helium, the founding of the journal 'Nature', and the Hill Observatory, now the Norman Lockyer Observatory. His is a complex tale because there is so much overlap between his interests and achievements. He was a friendly and intelligent man, keenly interested in the world around him, holding weekly get-togethers at his home in Wimbledon. I have come across more than one reference to his welcoming approach by having a pipe rack in which his friends could leave their pipes... ready for their next smoke and conversation get together. He was interested in a vast array of subjects and issues of the day and put a great deal of effort into these as well as his work in astronomy and subsequent research. A remarkable man.

Sir Norman was born on 17 May 1836 into a family interested in science and education and it is noted that the family home was witness to many electrical experiments. His early life was not easy, with his mother dying when he was 13, in 1849, after which he was sent away to live with a relative in Warwickshire, where he attended school at Kenilworth. He became proficient in Greek and Latin, progressing to become a language tutor, having travelled in order to improve his French and German.

He was subsequently appointed as a clerk in the War Office in London, rising early to take part in rifle practice (at the time there were concerns relating to possible war with France), visiting the British Museum Reading Room before work. It was here he met many interesting people, including one in particular who was studying the Sun. This latter piqued his interest and combined with being introduced to astronomy through looking through a neighbour's telescope, set him on a scientific path. He became an amateur astronomer and science journalist in addition to his work at the War Office.

His published articles led to a new appointment, as Secretary to the newly formed Devonshire Commission for the formulation of science policy. He established a new scientific journal, that of Nature and wrote many astronomy books. He was also appointed as a lecturer at the Normal School of Science where he produced astronomy teaching notes as well as using his own lantern slides. He was one of the pioneers in the use of visual aids in teaching. These slides are now held in the Norman Lockyer Observatory at Sidmouth.

He was made Professor of Astrophysics at the Normal School around 1887. The Normal School was renamed in 1890 as the Royal College of Science which then became part of what is now Imperial College. In 1902 he was made Director of Solar Physics at the College. He took part in numerous eclipse expeditions to Greece, Egypt, and Japan.

He developed a new technique for analysing the Sun's spectra and was a awarded a medal by the French government, this medal was jointly awarded to French astronomer, Pierre Janssen and Lockyer for their almost simultaneous solar flare research and the development of the new technique.

Spectroscopy was a growing science in the early 1880s, although the study of the solar spectrum had begun with Newton's study of the nature of light in the late 17th Century. Josef Fraunhofer was the first scientist to fully appreciate and study the significance of the dark divisions in the Sun's spectrum. The light from the Sun was passed through a narrow slit before being dispersed by a prism and took the form of a narrow band of light crossed with dark lines (Fraunhofer lines).

Sir Norman pressed for the need for an astrophysics observatory to be based at Cambridge in addition to the South Kensington centre. The entire Solar Physics Observatory was transferred totally to Cambridge in 1912. However, Sir Norman had wanted to continue with the research at South Kensington, with a separate school of solar physics at Cambridge so with the end of the solar physics research at South Kensington, Sir Norman made the decision to retire and establish a new observatory on land belonging to his second wife at Sidmouth.

The Helium Story

The journey to the discovery of Helium is a long one, beginning with Fraunhofer's research in 1814 using spectroscopic analysis of the colours emitted by gases. In 1868 a new thin yellow line near line D was identified but was not confirmed to be that of a new element, helium, until 1895 by the chemist Ramsay

Sir Norman began his solar research in 1866, concentrating on sunspots before moving on to examine the chromosphere. He suggested using a spectroscope to analyse the constituent elements of solar flares. During an eclipse expedition to India in 1868 he positively identified a new line on the spectrograph very close to Fraunhofer line D. The location of the new line was also confirmed by Norman Pogson, the Government Astronomer in India who became a good friend of Sir Norman.

While making observations of the Orion nebula at his observatory in Westgate on Sea in 1890, Sir Norman saw the same lines as those he had seen in the Sun. However, when he published his book 'Chemistry of the Sun', he omitted reference to helium because it had not been found on Earth and there were also still many other unidentified spectral emission lines.

But, during the 1890s helium came back into the picture of mainstream science following the research by a geologist in the United States, Hildebrand, whose experiments on uraninite yielded many unidentified emission lines. Unfortunately, his results were only circulated amongst other geologists which later prompted Sir Norman to call for a more interdisciplinary approach as this would have enabled the unknown element to have been discovered sooner.

In 1895 Lord Rayleigh (an eminent physicist) undertook geological chemistry research, identifying a new element and sent the results to Ramsay, an eminent chemist. Being aware of Hildebrand's research and results, Ramsay also undertook research into uraninite, identifying the element D3 which he sent in a tube to Sir Norman. Sir Norman analysed the sample as well as conducting his own experiments into uraninite, using a heating method. It was noted that the emission lines were the same as those found in stars and the Sun. The lines in the Sun were bright, whereas in stars they appeared dark. Other elements were also identified, and Sir Norman suggested a new order of elements was needed. These previously unknown elements were found deep within the Earth and provided a link to the stars as they had similar properties.

These same emission lines were also found in the Orion nebula by other scientists and Sir Norman himself took a photograph following his observations and spectroscopic analysis. Helium was the name given to the new emission line because the Sun (Greek helios) was where the element had first been discovered.

The efforts and achievements by Norman in his part of the discovery of helium were finally recognised when he was made Knight Commander of the Order of the Bath on 22 June 1897. He was one of those honoured by Queen Victoria on her Diamond Jubilee Honours List. His initial discovery of the new element was particularly special because it was first discovered outside of Earth.

Helium was later found to be a cause of some of the colour seen in aurorae and its liquid and refrigeration properties were shown to have huge potential industrially. In 1914 the element was also suggested for use in airships.

Thus, from the first discovery of that unknown new yellow line near line D, the way was paved for a whole new important group of elements which became known as the Noble Gases. The practical applications of helium now include common use in MRI scanners and cryogenics.

*****

THE NORMAN LOCKYER OBSERVATORY

Following the decision to close the solar physics observatory at South Kensington and combine solar physics research with astrophysical research at Cambridge, Sir Norman retired and established a new observatory on land owned by his second wife. He was elected President, Sir Francis Maclean Secretary, with HRH Duke of Connaught as President of the funding appeal. Observational work began in 1913, but war intervened, and it was not until 1916 that the site was incorporated as the Hill Observatory. It then continued as a working observatory with a paid astronomer until the 1950s. The lecture theatre is named after the final such person, Donald Barber. It then evolved into the Norman Lockyer Observatory managed by volunteers. Funding is now received from various sources including those from open events.

There is plenty of information available regarding Norman Lockyer, helium, spectroscopy and telescopes generally in the Library and amongst members at the NLO, in Sidmouth, Devon, as well as in the Exeter Central Library Stack and in the archives held at Exeter University.

The Observatory holds regular open days and evenings as well as occasional special events, details of which can be found on its website

www.normanlockyer.com

THE LOCKYER TELESCOPE

As a trained presenter of this historic telescope, I delight in asking people what the most important fact about it is.

It is not in a museum! That is the most important thing to remember and appreciate about this, and indeed all the other historic telescopes, at the NLO. This telescope contains the actual lens used by Norman Lockyer to make observations of the Orion nebula back in 1889. Similar historic telescopes are contained behind a glass screen in museums, keeping them safe but untouchable. There is nothing more powerful than touching an artefact to bring the past to life. It was built in 1871 by Thomas Cooke, manufacturers of scientific instruments, who made lenses for the film industry until recently. I like to think Sir Norman would be delighted to see the telescopes still in use and enjoyed by all.

OBSERVING EQUIPMENT

Naked Eyes

Underused resource. Ideal for learning the early stars to appear and the constellations. Recommended for familiarisation with the night sky.

With practice, averted vision can reveal wonders such as the Andromeda Galaxy. Overhead is the best place to start as this is less affected by atmospheric disturbance. Nearer the horizon, you are looking through more of the atmosphere which is also prone to light pollution. *Cost*: FREE (I personally find distance spectacles better than varifocals for observation work with or without equipment).

Planisphere

A planisphere comprises two discs. The rear disc shows all the stars visible from your location, while the front one has a window showing only part of the rear disc. Set the front disc to any date and time to see which stars are visible at that time and where they are. *Cost*: about £10.

Star map or atlas, Moon map or atlas

There are many resources available online or in print. *Cost*: variable.

Red Light Torch

These will allow you to find your way while your eyes are adjusting to the darkness. It can be a handheld one or a useful headtorch.

Binoculars

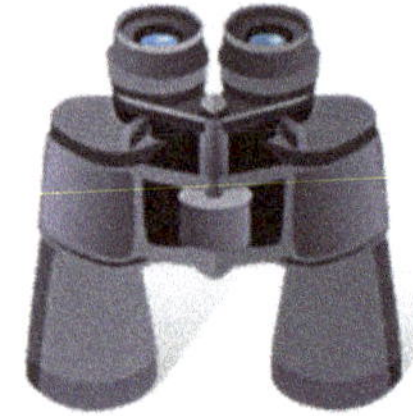

These are two refracting telescopes joined together; some have individual focussing. With their wide field of view, they are good for looking at Moon and distant star clusters.

Eye Relief: 20mm – the image focus is just beyond eye lens; this is essential for spectacle wearers, for whom rubber eye caps that fold back are helpful.

Exit pupil = diameter of light rays leaving the eye piece. This should not be greater than 7-8 mm; any excess light will not reach the retina. The diameter of the pupil when dark adjusted is around 5mm.

7x50mm are the minimum recommended; 10x50mm better. First figure is the magnification, and the latter refers to the aperture size – the greater the aperture, the more light can get in and thus a clearer image. Divide the aperture size by the magnification to find the exit pupil and 10x50s meet the optimum of 5mm.

The magnification size is less important as a small aperture gives a poor-quality image so greater magnification would not help.

Test by viewing against a straight edge which should be sharp. Once focussed, hold 10cm away and image should still be in focus.

Cost: Ranges from £20 if you're quick to catch an offer at Lidl, around £50-80 for a fairly decent pair (of 10x50mm) and onwards and upwards from there.

Lockyer Telescope
For a comparison, this is a 6¼" refractor for those familiar with the view through this telescope at the NLO.

Dobsonion
This is a Newtonian resting on ground with altazimuth mount, making it stable for observations.
Cost: upwards of £200.

Courtesy of Hannah Davies

Reflecting telescope

In these the image is upside down but not mirror reversed, giving a bright image. It is suitable for wider fields of view and has a lower magnification than

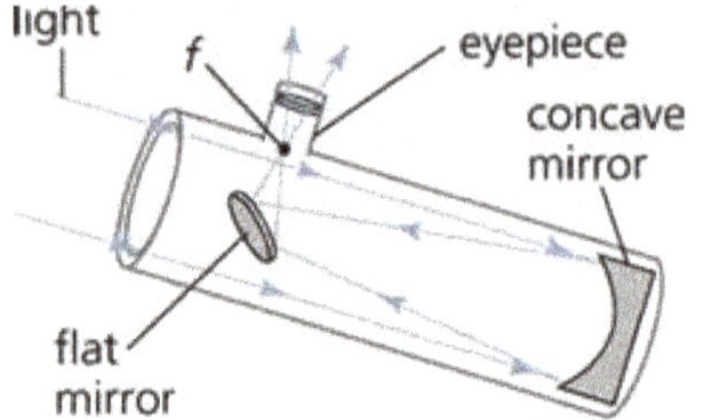

refractors. In a Newtonian reflector, the secondary mirror is flat which does not alter focal length. A Cassegrain reflector has a convex secondary mirror which increases the focal length.

Cost: around £180 upwards.

Refracting telescope

It is best for objects requiring long focal ratios and high magnification such as the Moon, planets and double stars.

It has a star diagonal (an angled mirror or prism that allows viewing from a direction that is perpendicular to the usual eyepiece axis) which allows more

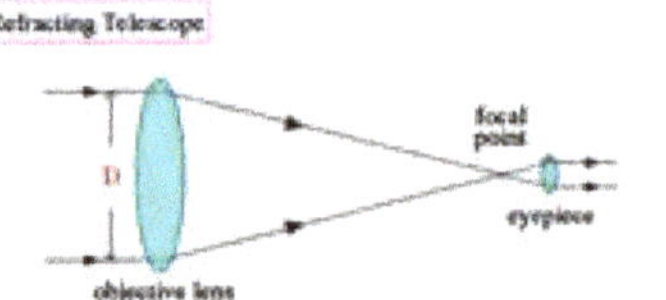

convenient and comfortable viewing.

The image is upside down and mirror reversed! An aperture of 75mm is the minimum required for observing.

Cost: £250 and upwards.

Catadioptric (Schmidt-Cassegrain)

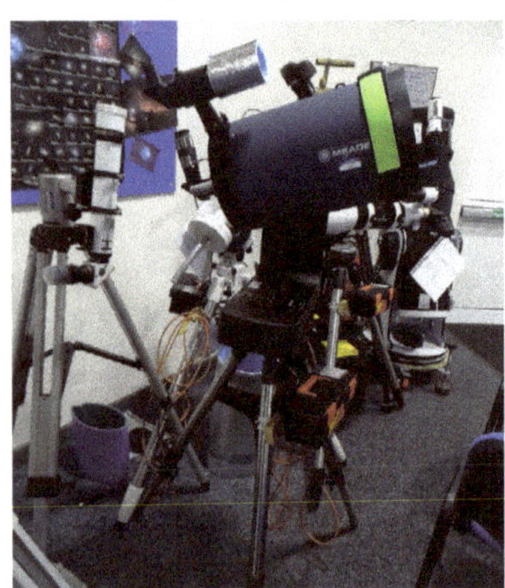

These have two mirrors and a corrective lens. There is a slight loss in quality due to secondary mirror. They have long focal ratios and are compact and portable. Ideal for the Moon and planets. *Cost:* around £2,000.

Other Equipment

Tripods
A necessary piece of equipment unless you have really strong and steady hands!

Mounts
Equatorial mounts have their axis parallel to the Earth's equator and are thus more convenient.

Altazimuth mounts are adjustable vertically and horizontally.

Lenses/eyepieces
These are categorised by their magnification ability.

Filters
Special filters are available for viewing the Moon, Venus and the Sun. The latter is essential for solar astronomy.

**** NEVER LOOK AT THE SUN DIRECTLY ****

Warm Clothing
With the best observing done in the winter months, do wrap up in lots of layers, wool jumpers, thick socks, hat, and prepare some hot drinks or better still, have a flask ready to keep you warm.

HOW TO USE A PLANISPHERE

These are widely available from shops and the internet. If you haven't got a smartphone app or access to a computer, a planisphere is a useful aid to learning the location of stars and planets (there is a table of planet locations for several years on the back). It will help you to find what stars you are looking at quite easily. It is also useful for armchair browsing before heading outside.

A planisphere comprises two discs. The rear disc shows all the stars visible from your location, while the front one has a window showing only part of the rear disc. Set the front disc to any date and time to see which stars are visible at that time and where they are.

To use it, face approximately south.
1. Hold the planisphere in front you like a reading book.
2. Rotate it so that 'south' on the planisphere is pointing towards the southern horizon you're facing.
3. Make a note of your zenith view ((overhead) and compare to the planisphere. Sometimes, it helps to hold the planisphere above your head so that it matches the sky.
4. Have fun!

OBSERVING LOG:
OBJECTS OBSERVED

Date and Time (GMT/ BST)	Object	Equipment Used	Weather Conditions	Notes

OBSERVING LOG:
OBJECTS OBSERVED

Date and Time (GMT/ BST)	Object	Equipment Used	Weather Conditions	Notes

GLOSSARY

Absolute magnitude
This refers to the brightness of a star if all stars were viewed from the same distance from Earth, i.e. 10 parsecs or 32.6 light years from Earth.

Almagest
Ptolemy's great work on astronomy written in AD127 – 151. contains a star catalogue and the motion of the Moon and planets. The rules he set out for calculating the future positions of the planets based on an Earth-centred universe were used for centuries.

Altitude
The angular distance of an object above or below the observer's horizon. An altitude of 0 degrees is at the horizon, with 90 degrees at the zenith (or point directly above the observer. Nadir is the point directly below an observer.)

Apogee
The point furthest away from Earth (apo = away from, geo = Earth).

Apparent magnitude
The apparent brightness of a celestial star or planet or other object as seen from an observer on earth.

Asterism
A pattern of stars either part of a constellation or a well-known pattern separate from any constellation.

AU
An astronomical unit is the unit of distance between the Earth and the Sun.

Averted vision
A technique to see distant, faint objects. It involves looking to the side of an object while trying to look at it. This allows more light to enter the eye. Keeping the object in the line of sight nearest the nose, will help to avoid it being in a blind spot of the eye.

Azimuth

The azimuth is the angle formed between a reference direction (in this example north) and a line from the observer to a point of interest projected on the same plane as the reference direction orthogonal to the zenith.

Binary star

This is a stellar system made up of two stars moving around a common centre of gravity.

Binoculars

These are two refracting telescopes joined together. With their wide field of view, they are good for looking at the Moon and distant star clusters.

Celestial equator

The projection into space of the Earth's equator; the circle on the celestial sphere.

Celestial sphere

An imaginary global shell surrounding the Earth at a distance. The ecliptic corresponds with the equator while the vertical lines of longitude equate to the values of right ascension with right ascension the astronomical equivalent of latitude.

Constellation

A recognised pattern of stars.

Craters

What are they? The effect of objects such as meteorites and planetesimals hitting the Moon. The rock is forced out upon impact pushing the rock sideways and upwards forming a crater wall, with the material falling back down and ploughing through the surface as rays. As rays get broken up over time, the craters with rays tend to be younger ones.

Declination

The celestial equivalent of latitude.

Degree

One degree is two Moon widths.

Double star

A star made up of two components either genuinely as in a binary star system or optically by chance (looks like a binary when viewed through binoculars).

Ecliptic

The apparent path the Sun takes around the Earth.

Exoplanet
An object orbiting a star other than our own Sun.

Globular cluster
These contain some of the oldest stars in our galaxy as they are formed from the original material of the galaxy, created from very few elements.

Gravity
From the Latin 'gravitas' meaning 'weight', it is a force which brings objects together.

HR diagram
The Hertzsprung-Russell diagram illustrates how stars fall into groups according to the stage in their life cycle. It was created by a Danish astronomer, Ejner Hertzsprung and an American astronomer, Henry Norris Russel.

IC catalogue
These Index Catalogues (there are two) are supplements to the NGC, both containing corrections to the NGC.

Latitude
The angle north or south of a reference point. (see Declination)

Longitude
A measurement east or west of a known reference point. (see Right Ascension)

Libration
A periodic wobble of a celestial body.

Light year
This is the distance light travels in a year.

The speed of light is 299,792 km/186,287.5 miles per second. Thus, you can work out the number of seconds in a year by multiplying the number of seconds in a minute (60) by the number of minutes in an hour (60), then multiply that number by the number of hours in a day (24), and multiply that by the number of days in a year (approximately 365.25).

So, we have:
60 x 60 = 3600 seconds in an hour
3600 x 24 = 86400 seconds in a day
86400 x 365.25 = 31,557,600 seconds in a year
Multiply by the distance per second.

A light year, therefore, is about 9,461,000,000,000,000 km/5,878,786,100,000 miles.

This is almost 9 trillion km/6 trillion miles.

Another way to think about this is to consider that 1 AU is the distance between the Earth and the Sun, on this basis light travels 63,240 AU.

The distance to the nearest star, Proxima Centauri is 4.3 light years while the distance to the Andromeda Galaxy is 2 million light years.

M/Messier
A catalogue of objects compiled by comet hunter, Charles Messier, in the 18th century. It is a list of objects that might be mistaken for comets.

Magnitude
This relates to the brightness of a star, the lower the magnitude, the brighter the object.

Apparent magnitude is the apparent brightness of a celestial star or planet, or other object as seen from an observer on earth.

Absolute magnitude is the brightness a star would have if it was viewed from the same distance as other stars.

Meteor Shower
What are meteor showers? When a comet nears the Sun, it becomes heated with the result that it sheds some of its outer hard coating leaving the classic tail behind it. The dust particles hang in space and Earth passes through this debris on an annual basis.

Moon
A separate body orbiting a planet.

Navigation stars
57 stars which are readily visible in early evening used to guide sailors and travellers.

Nebula
A cloud of gas and dust in space.

NGC (New General Catalogue)
A fairly comprehensive guide to deep sky objects compiled in 1888.

Open Cluster
The stars are more spaced out than in globular clusters and contain several hundred to several thousand stars a few light years across. They are relatively young formations.

Orbit

The Moon orbits Earth, and both Earth and the Moon orbit the Sun. Orbit is the path taken by one body around another.

Perigee

The closest point to Earth (peri = near, geo = Earth).

Planet

An object that does not give out light and has cleared the area around it of debris.

Planetary nebula

These are the remnants of old stars when they have exhausted their internal fuel supply, so named because they looked like planets to early astronomers.

Planisphere

A practical aid to locating stars

Precession

It is due to precession that the pole star changes over time and why it is thought the layout of the Pyramids at Giza reflect the layout of Orion which would have been above them at the time of building. Precession is a change in the orientation of the rotational axis of a rotating body – as can be seen in a gyroscope.

Red Dwarf

These are low mass, cooler stars.

Right Ascension

The celestial equivalent of longitude.

Rotation

The turning of a body on its axis.

Star

A ball of gas.

Telescope

An optical instrument using various combinations of mirrors and lenses to see distant objects. First used in the early 17th century, the telescope was developed out of spectacles for shipping purposes.

Tidal locking
This is when an astronomical body shows the same face to the body that it is orbiting, as is the case with the Earth and Moon.

Universe
A term used to describe the entire stellar and planetary system in which we live, indeed everything which exists; from the Latin 'universe' or 'uni versus' means one lot of turning.

Variable star
A star that varies in brightness.

Zenith
A point in the celestial sphere directly above the observer.

RESOURCES AND FURTHER READING

Details of activities, such as crater making, 3D model of Orion, making the Solar System to scale, suitable for teachers and group leaders can be found on www.karenhedges.co.uk

Further Reading

NASA websites

Norman Lockyer Observatory and website

'Stars of Orion' Oppenheimer

BIBLIOGRAPHY

Baker, Joanne, 50 Ideas you really need to know: Universe, 2010, Quercus

Chapman, Allan, Stargazers: Copernicus, Galileo, the telescope and the Church, 2014, Lion Hudson plc

Consolmagno, Guy and Davis, Dan M, Turn Left at Orion, 2011, Cambridge University Press

Lockyer, T Mary and Lockyer, Winifred L, Life and Work of Sir Norman Lockyer, Macmillan 1928

Meadows, A J, Science and Controversy: a biography of Sir Norman Lockyer, 1972, 2008, Macmillan

Moore, Patrick and Lintott, Chris, Astronomy for GCSE, 2001, Duckworth

Moore, Patrick, Brilliant Stars, 1996, Cassell

Moore, Patrick, Exploring the Night Sky with Binoculars, 1986, Cambridge University Press

Moore, Patrick, Guide to the Moon, 1976, 1977, Book Club Associates

Moore, Patrick, New Guide to the Planets, 1993, Sidgwick & Jackson

Scalzi, John, The Rough Guide to the Universe, 2008, Rough Guides Ltd

Tonkin, Stephen, Binocular Astronomy, 2014, Springer

ACKNOWLEDGEMENTS

I would like to thank the following people who have supported me in the production of this book and provided illustrations:

David Strange for the use of his photographs

Alan Green for the use of his photographs

Dr Alison Gray for her helpful explanations and artwork

John Meacham for his helpful explanations and artwork

Royston Williams for his diagrams and encouragement

Alan J Smith for his photograph (of the Last Quarter Moon)

A.E (Tony) Kingston for his photograph (of the Andromeda galaxy)

Additional Illustrations

NASA

Wikipedia Commons

ABOUT THE AUTHOR

From standing under a tree at primary school to witness a partial solar eclipse through the Space Age of the 1970s, through to being up and about at 3 am on a family holiday to view a meteor shower, Karen maintains her lifelong love of space, particularly the Moon and planets.

As a volunteer at the Norman Lockyer Observatory, Karen has many years' experience of helping both adults and children understand what it is they are observing. In the young people's section, she used a variety of slides, posters, and practical activities with which to engage her audience. She is a trained telescope presenter in the use and explanation of one of the historic telescopes and deems it a privilege to be able to do this. Her work as a teaching assistant has also informed the approach she takes in this guide to the stars and more.

She is an active STEM ambassador, promoting maths and science by giving talks at her local library and scout groups as well as the Observatory.
For an online radio programme, Karen records a monthly guide to the night sky, and this is available to listen to via:

http://www.mike-coles-travel.co.uk/newlife/